T0331265

Error Propagation in Environmental Modelling with GIS

Error Propagation in Environmental Modelling with GIS

GERARD B. M. HEUVELINK

University of Amsterdam

TAYLOR & FRANCIS
1798 – 1998

UK   Taylor & Francis Ltd, 1 Gunpowder Square, London EC4A 3DE
USA   Taylor & Francis Inc., 1900 Frost Road, Suite 101, Bristol, PA 19007

British Library Cataloguing-in-Publication Data

A catalogue record for this book is available from the British Library.

ISBN 0 7484 0743 X HB
ISBN 0 7484 0744 8 PB

Library of Congress Cataloging-Publication-Data are available

Cover design by Hybert Design and Type
Typeset in Times 10/12pt by Santype International Ltd, Salisbury, UK
Printed by T.J. International Ltd, Padstow, UK

Contents

Series introduction

Welcome to the *Research Monographs in Geographical Information Science*. The new *RMGIS* aim to provide a publication outlet of the highest quality for research in GIS which is longer than would normally be acceptable for publication in a journal. The series will include single and multiple author research monographs, possibly based upon PhD theses and the like, and special collections of thematic papers.

The need

We believe that there is a need, from the point of view both of readers (researchers and practitioners) and of authors, for longer treatments of subjects related to GIS than are widely available currently. We feel that the value of much research is actually devalued by being broken up into separate articles for publication in journals. At the same time, we realise that many career decisions are based on publication records, and that peer review plays an important part in that process. Therefore a named editorial board has been appointed to support the series, and advice will be sought from them on all submissions.

Successful submissions will focus on a single theme of interest to the GIS community, and treat it in depth giving full proofs, methodological procedures or code where appropriate to help the reader appreciate the utility of the work in the monograph. No area of interest in GIS is excluded although material should demonstrably advance thinking and understanding in spatial information science. Theoretical, technical and application-oriented approaches are all welcomed.

The medium

In the first instance the majority of monographs will be in the form of a traditional textbook, but, in a changing world of publishing, we actively encourage publication on CD-ROM, the placing of supporting material on web sites, or publication of programs and of data. No form of dissemination is discounted, and prospective authors are invited to suggest whatever primary form of publication and support material they think is appropriate.

The editorial board

The monograph series is supported by an editorial board. Every monograph proposal is sent to all members of the board which includes Ralf Bill, Antonio Camara, Joseph Ferreira, Pip Forer, Andrew Frank, Gail Kucera, Peter van Oostrom and Enrico Puppo. These people have been invited for their experience in the field, of monograph writing, and their geographic and subject diversity. With particular monographs members may also be involved later in the process.

Future submissions

Anyone who is interested in preparing a research monograph should contact either of the editors. Advice on how to proceed will be available from them, and is treated on a case by case basis.

For now we hope that you will find this, the first in the series, a worthwhile addition to your GIS bookshelf, and that you may be inspired to submit one too.

Series editors

PETER FISHER

Department of Geography,
University of Leicester,
Leicester LE1 7RH, UK
Telephone (+44) (0) 116 252 3839.
Fax (+44) (0) 116 252 3854.
Email pffl@le.ac.uk

JONATHAN RAPER

Department of Geography,
Birkbeck College, 7–15 Gresse Street,
London W1P 1PA, UK
Telephone (+44) (0) 171 631 6457.
Fax (+44) (0) 171 631 6498.
Email j.raper@geog.bbk.ac.uk

Foreword

It is a great honour to be asked to write this brief foreword to Gerard Heuvelink's monograph. As Gerard notes in his Preface, the monograph is the culmination of a decade of work, and it brings together and extends results that have been published in a number of forms over that period. Gerard has made an outstanding contribution to the field over that period, and his work is far and away the best and most comprehensive attempt to formalise and offer solutions to the difficult problem of error propagation in GIS. Publication of this monograph is very timely, as it makes all of this work accessible in one source, and brings it to the attention of a wide and multidisciplinary audience in a format that is both rigorous and relatively easy to implement.

Error propagation in GIS may appear to the uninitiated as an obscure, abstruse topic within a relatively obscure field – so why is it so important, and why does it deserve attention in the form of this monograph? Aren't computers accurate, and isn't everything that emerges from them perfectly true? We have made great progress in the past decade or so in extending computer applications into the domain of the vague and imprecise, and computers are now used to help make decisions under conditions of uncertainty; computers armed with tools such as the World Wide Web are far from the exact number-crunchers that defined the computing world of thirty years ago. But in the case of geographic information the problem of unreasonable expectations is made worse by two additional factors. First, uncertainty is endemic in geographic information, since it is virtually impossible to record and represent the true complexity of the Earth's surface in digital form. But perhaps more important is the fact that the previous technology – in this case, the paper map – was also guilty of an exaggerated sense of its own precision. Somehow, we have grown used to the idea of precisely drawn contours on topographic maps, and precise boundaries on soil maps, even though we know on reflection or closer examination that their true accuracy is much lower. Somehow we accept this pretence of accuracy, and have developed a technology of geographic information systems that takes and analyses that map information at face value, without questioning its true accuracy. We are only now coming to realise that much of what passes for accurate analysis in GIS is in fact highly uncertain, and that decisions and regulations that depend on analysis can be flawed. Pity the poor GIS analyst required to defend in court the results of a GIS analysis that did not adequately reflect the uncertainty in its data, against a clever and well-informed lawyer. The only answer is a new generation of accuracy-aware GIS, using the techniques and following the examples discussed in this monograph.

Gerard uses the term 'error' in the statistical sense of variation, and explains why the problem is rather more general than the term might imply. It is important to realise that the methods discussed here can be applied not only to the problem of error in measurement, but also to the uncertainties introduced by inadequate definitions, excessively coarse scale, insufficient samples, and many other sources. The definition of accuracy used here – the "difference between reality and our representation of reality" – contains that all-important word 'our'. If 'we' are not all of one mind – if different observers might make slightly different maps of the soils in the same area – the variation that results is a form of error in the sense of this monograph.

Error analysis in GIS would be much easier if map-making were more like scientific measurement; for example, if a digital elevation model (DEM) were actually built by going out in the field and measuring directly the elevation at each point in the DEM's grid. In reality, the construction of a geographic database is often a complex process involving many different people, different sets of measurements, stages of interpretation and manipulation that more or less completely confound the relationship between original measurements and final database contents. The process is often undocumented, and only in rare circumstances is it possible to develop a formal analysis of error. Moreover, the response of the environment to uncertain inputs is often highly nonlinear. A small error or uncertainty in a digital elevation model can make a dramatic difference to the calculation of the area visible from a point, or the geometry of a drainage network. The net result is that our intuitive understanding of the connection between uncertain input and uncertain output of an analysis or a model is extremely limited. It can be close to impossible to guess how much impact a given input uncertainty will have, without the aid of error propagation methods.

So what next, and where do we go from here? Gerard's monograph includes many tools, and the means to implement them is available in packages like ADAM. It includes excellent examples of the use of the methods in classic GIS analyses. It is important to note, however, that while GIS and environmental modelling overlap, they are not the same, and that the methods are potentially applicable to a much wider domain. Finite-element methods are often used for modelling environmental systems without reference to GIS; and for modelling in spaces that are not normally thought of as geographic. The tools would be much more widely accessible if they could be incorporated into the mainstream commercial GISs, and into software environments that are also commonly used to support environmental modelling, such as the statistical and mathematical packages. Hopefully, this work will also stimulate further research on the other forms of geographic data, particularly discrete object representations, and the particular uncertainties to which they are subject.

MICHAEL F. GOODCHILD
University of California

Preface

This monograph is the product of almost a decade of research that I dedicated to the problem of error propagation in GIS. It all began in 1988, when Peter Burrough offered me a job as a research assistant at the University of Utrecht. He must have taken a (calculated?) risk by employing an applied mathematician, who associated geography with memorising the names of cities and provinces and who was completely unaware of the mere existence of geographical information systems. However, with the help of my colleagues I soon mastered the basics of physical geography and GIS (and that's about the level that I stayed on to this day), and my background in mathematical statistics came more and more to my advantage. In any case, it was my strong belief that the theory on error propagation in GIS would only advance if it had a firm statistical basis. I put these ideas into practice and five years later I rounded my work in Utrecht off with my PhD thesis, which also appeared as an issue of the *Netherlands Geographical Studies*.

My work did not go unnoticed by Peter Fisher, editor of the *International Journal of GIS*, who proposed to have my thesis published as a research monograph by Taylor & Francis. I gladly accepted the offer and the result lies now in front of you. Clearly, the core of this monograph is taken from my thesis. However, I took the opportunity to update the work with recent developments and to include work that I have more recently been doing at the University of Amsterdam. This means that in this monograph more attention is paid to the assessment of input error. This is important in my view because an error propagation analysis can only yield sensible results once the uncertainty of the inputs to the analysis are correctly identified.

This monograph is addressed to geographers and environmental scientists with a quantitative attitude to the profession. It may also be used by advanced students with a particular interest in the subject. Some may find the statistical and geostatistical parts of the book somewhat hard to digest, but I do hope that they will find the motivation to persevere. If it is of any help: I did not include the statistics to make things more complicated than they need be, but because I believe that we need a sound footing of the theory to truly advance.

Very many people have contributed to the realisation of this monograph, but two must in particular be named. Without Peter Burrough this monograph would simply not have existed. Not only did Peter initiate the research and pave the way for me with his own work on error propagation in GIS, as my supervisor he also played a crucial role in the successful completion of my PhD research. Through his involvement with my PhD research, Cees Wesseling has also made a major contribution to this work. Of all his skills, Cees' clear view on how error propagation functionality should be incorporated into a GIS is just one that I benefited from. I sincerely thank Peter, Cees and the many others who have helped me in writing this book.

GERARD HEUVELINK
Amsterdam

Glossary

Attribute   Property of a geographic object or location

Coefficient   Coefficient of a computational model

Computational model   Mathematical representation of reality; computational models can be classified into logical, empirical and conceptual (physically-based) models

Error   Difference between reality and our representation of reality; it includes not only 'mistakes' or 'faults' but also the statistical concept of 'variation'

Error model   Stochastic model to define the error in quantitative spatial attributes

Error propagation   Occurs when the errors of the input attributes to a GIS operation cause errors in the output of the operation

Identification   Estimation of the parameters of the error model

Input error   Error in the inputs to a GIS operation

Model error   Error of a computational model

Operation   Allows one to derive new attributes from attributes already held in the GIS database; operations can be classified into local (i.e. point and neighbourhood) and global operations

Output error   Error in the output of a GIS operation

Parameter   Parameter of the error model or of a stochastic model of spatial variation

Simulation   Refers to stochastic simulation; generating a realisation of a random variable or random field

Spatial modelling   The use of GIS operations to evaluate computational models

Spatial variability   Refers to the small-scale and large-scale spatial variations that are present in a spatial attribute; exists regardless of how the attribute is mapped

Support   Size, shape and orientation of field observations or of the attributes stored in the GIS database

Uncertainty   Synonymous with error; the uncertainty of a spatial attribute is affected by the spatial variability and the mapping procedure used

CMSV   Continuous model of spatial variation; version of the GMSV that assumes that spatial variation is gradual

DMSV   Discrete model of spatial variation; version of the GMSV that assumes that major jumps in attribute values take place at the boundaries of mapping units

GMSV   General model of spatial variation; a stochastic model to describe the spatial variability of an attribute

MDL   Model description language; programming language in which the user formulates an error propagation problem

MMSV   Mixed model of spatial variation; version of the GMSV that is in between the DMSV and CMSV and can handle both abrupt and gradual changes

List of symbols

The list of symbols is not exhaustive but it contains the symbols that are used more frequently in the text.

$a(\cdot)$	Realisation of $A(\cdot)$, unknown true value of spatial attribute
$b(\cdot)$	Mapped representation of $A(\cdot)$
$f(\cdot)$	Probability density function
$g(\cdot)$	GIS operation
h	Distance vector, $h = x - x'$
i, j, k, l	Indices
m	Number of inputs to a GIS operation
m_u	Sample mean of U using the Monte Carlo method
n	Number of observations in D
n_k	Number of observations in unit D_k
p_v	v-percentile of U, probability that U is less than v equals p_v
q_α	α-quantile of U, probability that U is less than q_α equals α
s_u^2	Sample variance of U using the Monte Carlo method
$v(\cdot)$	Realisation of $V(\cdot)$, unknown error in spatial attribute
x, x'	Spatial locations, elements of D
$A(\cdot)$	Spatial attribute with uncertainty, random field defined on D
$A(x)$	Value of $A(\cdot)$ at location $x \in D$, random variable
C_0	Variance of residual $\varepsilon(\cdot)$ with DMSV
$C_Z(\cdot)$	Autocovariance function of second order stationary $Z(\cdot)$
$C_\varepsilon(\cdot)$	Autocovariance function of second order stationary $\varepsilon(\cdot)$ with MMSV
D	Domain of interest, $D \subset \mathbb{R}^n$
$D_k, k = 1, \dots, K$	Partitioning of the domain D
$F(\cdot)$	Cumulative distribution function

N	Number of Monte Carlo runs
$R(\cdot,\cdot)$	Autocovariance function of $V(\cdot)$
$R_{ij}(\cdot,\cdot)$	Cross-covariance function of $V_i(\cdot)$ and $V_j(\cdot)$
$U(\cdot)$	Output map of GIS operation
$V(\cdot)$	Error random field, $V(\cdot) = A(\cdot) - b(\cdot)$
$Z(\cdot)$	Spatial attribute with spatial variability
$Z_C^*(\cdot)$	Predictor of $Z(\cdot)$ under CMSV
$Z_D^*(\cdot)$	Predictor of $Z(\cdot)$ under DMSV
$Z_M^*(\cdot)$	Predictor of $Z(\cdot)$ under MMSV
$\beta(k)$	Mean of $Z(x)$ for any $x \in D_k$, parameter of the DMSV and MMSV
$\gamma_Z(\cdot)$	Variogram of $Z(\cdot)$
$\gamma_\varepsilon(\cdot)$	Variogram of $\varepsilon(\cdot)$
$\varepsilon(\cdot)$	Residual of attribute with spatial variation, $\varepsilon(\cdot) = Z(\cdot) - \mu(\cdot)$
$\zeta(\cdot)$	Mean of $U(\cdot)$
$\kappa_i, \lambda_i, \nu_j$	Kriging weights
$\mu(\cdot)$	Mean of $Z(\cdot)$
$\xi(\cdot)$	Mean of $V(\cdot)$
$\rho(\cdot,\cdot)$	Autocorrelation function of $V(\cdot)$
$\rho_{ij}(\cdot,\cdot)$	Cross-correlation function of $V_i(\cdot)$ and $V_j(\cdot)$
$\sigma^2(\cdot)$	Variance of $V(\cdot)$
$\sigma_C^2(\cdot)$	Variance of prediction error with CMSV
$\sigma_D^2(\cdot)$	Variance of prediction error with DMSV
$\sigma_M^2(\cdot)$	Variance of prediction error with MMSV
$\tau^2(\cdot)$	Variance of $U(\cdot)$

CHAPTER ONE

Introduction

During the last twenty years the handling of geographic information has undergone a major change due to the rapid emergence of geographical information systems. A geographical information system (GIS) can be conveniently defined as a software package for the storage, analysis and presentation of geographical information. A GIS thus provides the user with a set of tools for collecting, storing, retrieving at will, transforming and displaying spatial data from the real world for a particular set of purposes (Burrough, 1986). GISs have quickly become an indispensable tool for efficient management of spatial data and are now routinely being employed by academics, governmental agencies and private industry (Goodchild *et al.*, 1993; Kovar and Nachtnebel, 1993; Longley *et al.*, 1998; Burrough and McDonnell, 1998).

Geographic information can be represented in a GIS by two fundamentally different approaches. These are the *object* and *field* approaches (Goodchild, 1989, 1992). In the object approach, the GIS represents the world as being populated by simple objects – points, lines and areas – that are characterised by their geometrical and topological properties and by their non-spatial *attribute* values. Typical examples of spatial objects are buildings, roads, pipelines and parcels. Attribute values of these objects may then refer to the number of stories of a building, the width of a road or the diameter of a pipeline, or to the size and ownership of a parcel. In contrast, in the field approach the GIS represents the world simply as fields of attribute data, without defining abstract geographical objects. Examples of such fields are elevation, the infiltration capacity of the soil and the concentration of pollutants in the soil. Fields only provide the value of an attribute at any location and do not link these values with any recognised objects. The distinction between objects and fields is often associated with the distinction between vector and raster GIS, but in fact it entails much more than only the technical aspects of representing

1

and storing geographic information (Goodchild, 1989, 1992; Burrough, 1992a).

GIS operations

There are many ways in which information can be obtained from a GIS database. The most common way is simple data retrieval, such as finding the locations of all borings on a refuse tip where the level of pollution exceeds a critical value, or delineating an area where elevation is below sea-level. The use of structured query language (SQL) extends simple data retrieval to more complex queries, such as can be used for identifying feasible sites for dumping nuclear waste (Openshaw *et al.*, 1991). However, particularly in the earth and environmental sciences, GIS is also used to derive new attributes from attributes already held in the GIS database. For example, elevation data in the form of a digital elevation model (DEM) can be used to derive maps of slope and aspect. Digital maps of soil type and slope can be combined with information about soil fertility and moisture supply to yield maps of suitability for growing maize (Burrough, 1986).

In all these transformations the GIS is used to create new data from existing data. The data transformation is expressed as a mathematical function by which new attributes are derived. There are many basic types of function used for this kind of derivation and they are often provided as standard functions or *operations* in many GISs, under the name of 'cartographic algebra' (Burrough, 1986; Tomlin, 1990).

GIS operations can be grouped into three major classes of spatial extent. In *point operations*, the new attribute value of an object or at a field location is computed from other attributes relating to the same object or location. This is probably the most commonly used operation. In *neighbourhood operations*, the new attribute value is derived from the values of the same attribute, alone or in combination with other attributes, drawn from a window or kernel area surrounding the object or location of interest. This kind of operation is commonly used in raster GIS to filter surfaces to remove noise, to compute maps of slope and aspect from a DEM, or for edge detection in remotely sensed images. Point and neighbourhood operations together are referred to as *local operations*. In *global operations*, the value of an attribute is derived from an analysis of data that cover a larger area than a simple kernel. An example is the analysis of a DEM to find all locations in an area that are visible from the top of a watchtower. Another example comes from catchment hydrology, where the channel flow at the outlet of a catchment depends on precipitation, land form, land use and soil characteristics of the entire catchment.

Spatial modelling with GIS

In practice, many GIS operations are used in sequence to compute an attribute that is the result of a *computational model*. A computational model is understood here as a simplified mathematical representation of reality. For instance, the channel flow at the outlet of the catchment is computed after the relevant hydrological processes have been translated into mathematical equations – thus, after reality has been approximated by a suitable computational model. Using GIS for the evaluation of computational models is identified here by the term *spatial modelling* with GIS.

Computational models can be classified into *logical, empirical* and *conceptual* models. A logical model computes a new attribute by applying simple logical 'rules' to input attributes. For instance, using methods of logical intersection within a GIS, a user can create a map of erosion hazard by overlaying maps of soil texture, vegetation cover and slope (De Roo, 1993; Heuvelink and Burrough, 1993). Empirical models are based on an empirical understanding of the links between model input and output. These models rely on regression to estimate their coefficients and are often only applicable to the area in which they are derived. Many so-called pedotransfer functions used in land evaluation are empirical models (Cosby *et al.*, 1984; Bouma, 1989; Vereecken *et al.*, 1989). Conceptual models are based on a fundamental understanding – in physical terms – of the process being modelled and, in principle, can be applied generally. These models are also referred to as physically-based models. The coefficients of conceptual models are not empirical but refer to real physical properties. Examples of conceptual models used in the environmental sciences are distributed dynamic catchment models (Abbott *et al.*, 1986; De Roo, 1993; Binley *et al.*, 1997), models of physical and chemical soil processes (Tietema and Verstraten, 1992; Bouten, 1993; Finke *et al.*, 1998) and groundwater flow models (McDonald and Harbaugh, 1984). In practice, many of the models used in the environmental sciences contain both conceptual and empirical elements. For example, this is the case for many distributed soil erosion models (De Roo, 1993) and for crop yield models (Dumanski and Onofrei, 1989; Van Diepen *et al.*, 1991; Van Lanen *et al.*, 1992).

Relatively simple computational models can be executed directly in the GIS using the tools of cartographic algebra (Burrough, 1986; Tomlin, 1990; Burrough and McDonnell, 1998). Recent developments show that in principle it is also possible to build and implement more complex physically-based models within the GIS (Goodchild *et al.*, 1993; Kovar and Nachtnebel, 1993; Van Deursen, 1995; Wesseling *et al.*, 1996). However, in many situations it still is more efficient to run these models outside the GIS, with a link to the GIS database for retrieval of input data and to store results (De Roo *et al.*, 1989). The model results can then be displayed by the GIS or used as input to a new GIS operation.

To date, most work on spatial modelling with GIS has been concentrated on the business of deriving computational models that operate on spatial data, on the building of large spatial databases, and on linking computational models with the GIS. The large amount of effort spent on these issues clearly demonstrates the importance and value of spatial modelling with GIS.

However, there is a potential danger in GIS that has long received too little attention. This concerns the issue of data quality and how errors in spatial attributes propagate through GIS operations.

1.1 Error sources in GIS data

The problem of spatial data quality is obvious because no map stored in a GIS is completely error-free. Here the word 'error' is used in its widest sense to include not only 'mistakes' or 'faults', but also to include the statistical concept of error meaning 'variation' (Burrough, 1986). The data stored in a GIS have been collected in the field, have been classified, interpreted, estimated intuitively, and so contain a certain amount of error. Errors also derive from measurement error, from spatial and temporal variation and from mistakes in data entry.

When spatial data are entered into a GIS, this is often done by means of digitising a paper map. Clearly, errors are introduced during the transfer of the source map to the digital database (Maffini et al., 1989; Bolstad et al., 1990; Dunn et al., 1990; Keefer et al., 1991), but more important is that the uncertainties contained in the map are duplicated in the GIS (Muller, 1987; Goodchild et al., 1992). Most conventional maps of soil, geology, vegetation or land use make use of polygons to represent spatial attributes, meaning that areas of equal value are separated by crisp boundaries. However, particularly for environmental data, polygons are not very suitable to represent the real world situation. In reality, boundaries are often gradual, and mapping units are rarely truly homogeneous (Burrough, 1986, p. 120, 1993; Goodchild, 1989; Voltz and Webster, 1990; Oberthür et al., 1996). Therefore, the restriction to a polygonal representation inevitably causes the map to differ from reality. In addition to this, paper maps suffer from other sources of error as well, such as from reproduction, deformation and generalisation errors. Detailed discussions of the many different sources of error are given by Burrough (1986), Muller (1987), Veregin (1989a), Thapa and Bossler (1992) and Burrough and McDonnell (1998).

Not all maps stored in a GIS originate from paper maps. In the earth sciences, digital maps are often created by interpolation from point observations (Burrough and McDonnell, 1998). However, these maps also contain errors; the prevailing ones being caused by measurement and interpolation error. In many cases these errors can be quantified by employing geostatistical interpolation (Journel and Huijbregts, 1978; Webster and Oliver, 1990; Cressie, 1991). Useful research has also been done to assess and analyse the error in DEMs (e.g. Theobald, 1989; Fisher, 1990; Carter, 1992; Bolstad and Stowe, 1994; Hunter and Goodchild, 1997).

It is thus fair to conclude that virtually all data stored in a GIS are, at least to some extent, contaminated by error (Goodchild and Gopal, 1989, p. xii;

Openshaw, 1989; Heuvelink and Burrough, 1993; Burrough and McDonnell, 1998). What is not so well understood is how these errors contribute to uncertainties in the results of GIS operations and computational models.

1.2 The propagation of errors through GIS operations

When maps that are stored in a GIS database are used as input to a GIS operation, the errors in the input will propagate to the output of the operation. This is because the resulting output is a function of the input values, and inaccurate input values automatically affect the computed result (Heuvelink *et al.*, 1989). Therefore the output may not be sufficiently reliable for correct conclusions to be drawn from it. Moreover, the error propagation continues when the output from one operation is used as input to an ensuing operation. Consequently, when no record is kept of the accuracy of intermediate results, it becomes extremely difficult to evaluate the true accuracy of the final result.

The problem is even more complex because *input error* is not the only error that propagates through GIS operations. I mentioned before that in many cases the GIS operation is in effect a computational model, such as a model of crop growth or surface runoff. Since the computational model is merely an approximation of reality, it also contains errors, here referred to as *model error*. Input and model error both affect the output and both must therefore be accounted for.

Although users may be aware that errors propagate through their analyses, in practice they rarely pay attention to this problem. Perhaps experienced users know that the quality of their data is not reflected by the quality of the graphical output of the GIS, but they cannot truly benefit from this knowledge because the uncertainty of their data still remains unknown. Hardly any GIS currently in use can present the user with information about the confidence limits that should be associated with the results of an analysis (Burrough, 1992b; Lanter and Veregin, 1992; Klinkenberg and Joy, 1994; Burrough and McDonnell, 1998). In fact, most GISs do not even carry information about the uncertainty of the source maps in the spatial database. An important reason for this is that we still do not have a single, generally accepted theory for handling error and error propagation in GIS (Chrisman, 1989; Goodchild and Gopal, 1989, p. xiv; Lanter and Veregin, 1992; Forier and Canters, 1996; Hunter and Goodchild, 1997; Kiiveri, 1997).

1.3 Objectives of this study

There are many different aspects of the problem of error and error propagation in GIS. Therefore, any research that aims at obtaining concrete solutions should necessarily start with selecting a particular segment within the whole research area of error handling and error propagation in GIS. In this research,

I have chosen solely to restrict myself to the propagation of quantitative attribute errors in logical, empirical and conceptual models that are used in, or with GIS. This means that neither *positional error* nor the propagation of qualitative or *categorical errors* are dealt with here. This does not mean that these are not important issues, but they appear to be of minor practical importance for many applications in the environmental sciences.

Since this monograph only deals with the problem of attribute accuracy, it ignores the remaining five items from the list of six standard error components, which are positional accuracy, lineage, logical consistency, completeness and temporal accuracy (DCDSTF, 1988; Guptill and Morrison, 1995). However, these other five components are mainly important when one employs the object approach towards spatial data, whereas many of the problems in the environmental sciences can more suitably be dealt with using the field approach. Errors in objects are difficult to handle, and much research has been, and continues to be, committed to this (Chrisman, 1989; Griffith, 1989; Lemmens, 1991; Caspary and Scheuring, 1992; Edwards and Lowell, 1996; Hunter and Goodchild, 1996; Stanislawski *et al.*, 1996; Kiiveri, 1997).

This research also excludes the analysis of qualitative or categorical errors in GIS. Research on error propagation with categorical spatial data can be found in Newcomer and Szajgin (1984), Veregin (1989b), Lodwick *et al.* (1990) and Lanter and Veregin (1992). A first step towards an error theory for categorical spatial data that includes spatial dependence was given by Goodchild *et al.* (1992). Another valuable approach to modelling categorical spatial error originates from non-parametric geostatistics (Bierkens and Burrough, 1993a).

A list of nine research questions

The aim of this research is to develop a methodology for handling error and error propagation in quantitative spatial and environmental modelling with GIS. Among others, the study should provide answers to the following questions:

1 What is a suitable definition of error for quantitative spatial attributes?

2 Given this definition of error, how should the error for a particular input attribute to a GIS operation be identified?

3 Which techniques can be used to analyse the propagation of errors through local and global GIS operations?

4 What are the advantages and disadvantages of the various error propagation techniques?

5 How do these error propagation techniques perform when they are applied to practical problems?

6 When the GIS operation is in effect a computational model, how should model error be incorporated in the error propagation analysis?

7 When multiple error sources cause error in the output of a GIS operation, how can the relative contributions of the error sources be determined?

8 If the error in the output exceeds a critical level, what action can best be undertaken to improve accuracy? Conversely, if the quality of the output exceeds requirements by far, how can savings in data collection and processing best be made?

9 How should error propagation techniques be implemented in a GIS?

These and other questions will be considered during the course of this study.

CHAPTER TWO

Definition and identification of an error model for quantitative spatial attributes

2.1 A first look at quantitative errors

When we talk about error, everybody has an intuitive understanding of what is meant by it, but is it really clear? It is worthwhile to examine this matter more closely, and to define clearly what we mean by 'error'. Here, error will be defined as the *difference between reality and our representation of reality*. Although this definition does not always make sense (for instance, there are huge irrelevant differences between a city and a city map, even when the map is 'error-free'), it is appropriate for describing quantitative errors. For instance, if the nitrate concentration of the shallow groundwater at some location equals 68.6 g/m^3, while according to the map it is 62.9 g/m^3, then there will be no disagreement that in this case the error is $68.6 - 62.9 = 5.7$ g/m^3. Generalising this example, let the true value of a spatial attribute at some location x be $a(x)$, and let the representation of it be $b(x)$. Then, according to the definition, the error $v(x)$ at x is simply the arithmetical difference $v(x) = a(x) - b(x)$.

We consider the situation in which the true value $a(x)$ is unknown, because if it were known, then error could simply be eliminated by assigning $a(x)$ to $b(x)$. What is known exactly is the representation $b(x)$, because this is the estimate for $a(x)$ that is available from the map. The error $v(x)$ is also not known exactly, but we should have some idea about the range or distribution of values that it is likely to take. For instance, we may know that the chances are equal that $v(x)$ is positive or negative, or we may be 95 per cent confident that $v(x)$ lies within a given interval.

Knowledge about the error $v(x)$ is thus limited to specifying a range or distribution of possible values. This type of information can best be conveyed by representing it as a *random variable $V(x)$*. Note that a notation with capitals

9

is introduced here, to distinguish random variables from deterministic variables. Typically, a random variable is associated with the outcome of a probabilistic experiment, such as the throw of a die or the number drawn in a
lottery. But a random variable is equally suited to model the uncertainty
attached to an attribute that is not the outcome of a probabilistic experiment.
For instance, since we do not know the true nitrate concentration of the
shallow groundwater, we may think that it is a value drawn from a large set of
values that surround the estimated value of 62.9 g/m^3. Although we are aware
that the attribute has only one fixed, deterministic value $a(x)$, our *uncertainty*
about $a(x)$ allows us to treat it as the outcome of some random mechanism
$A(x)$. We must then proceed by specifying the rules of this random mechanism,
by saying how likely each possible outcome is. This will be done more for-
mally in the next section.

2.2   Definition of the error model

Consider a quantitative spatial attribute $A(\cdot) \equiv \{A(x) \mid x \in D\}$ that is defined
on the domain of interest D in n-dimensional space $\mathbb{R}^n$. In most cases $n = 2$ or
perhaps $n = 3$ but there is no real restriction. Refer to the value of $A(\cdot)$ at
some location $x \in D$ as $A(x)$. The *error model* introduced in the previous
section thus becomes:

$$A(x) = b(x) + V(x) \quad \text{for all } x \in D \tag{2.1}$$

where $A(x)$ and $V(x)$ are random variables and $b(x)$ is a deterministic variable.
Note that the word 'model' as it is used here refers to a stochastic or statistical
representation of attribute error, and should not be confused with the computational 'models' discussed in the previous chapter. Note also that $A(\cdot)$ and
$V(\cdot)$ are not random variables but random fields, in the geostatistical literature also termed random functions (Journel and Huijbregts, 1978; Cressie,
1991).

Error at a single location

Let us first consider the error at one location x only. Denote the mean and
variance of $V(x)$ by $E[V(x)] = \xi(x)$ and $\text{var}(V(x)) = \sigma^2(x)$. The mean $\xi(x)$ is
often referred to as the systematic error or bias, because it says how much $b(x)$
systematically differs from $A(x)$. The standard deviation $\sigma(x)$ of $V(x)$ character-
ises the non-systematic, random component of the error $V(x)$. Let $F_{V(x)}(\cdot)$ be
the cumulative distribution function of $V(x)$. When $V(x)$ is (absolutely) continuously distributed, then the probability density function of $V(x)$ exists, and it is
denoted here by $f_{V(x)}(\cdot)$. It is common practice to refer to both $F_{V(x)}(\cdot)$ and
$f_{V(x)}(\cdot)$ as the distribution of $V(x)$. Note that (2.1) implies that the attribute
$A(x)$ and error $V(x)$ have the same distribution, except for a shift in the mean.

In standard error analysis, it is usually assumed that errors follow the normal (Gaussian) distribution (Parratt, 1961; Topping, 1962; Taylor, 1982). The reason for this lies mainly in the central limit theorem, which roughly states that the averaging of a sufficiently large number of random variables yields a normal distribution, no matter what the distribution of the individual random variables is (Vanmarcke, 1983, p. 56; Casella and Berger, 1990). So when the error $V(x)$ can be seen as the superposition of various smaller error sources, then it is reasonable to assume that it is normally distributed. An additional, more practical reason for invoking the normal assumption is that normal theory is thoroughly worked out and often simple in form (Casella and Berger, 1990, p. 103; Helstrom, 1991). However, this does not mean that it is always safe to assume that $V(x)$ is normally distributed. It may sometimes be more appropriate to assume that $V(x)$ has another distribution type, such as the uniform, exponential or lognormal distribution. In fact, many attributes studied in the earth sciences are skewed and follow a lognormal distribution (Journel and Huijbregts, 1978; Davis, 1986; Isaaks and Srivastava, 1989; Webster and Oliver, 1990).

Spatial extension

Although a complete characterisation of the error random field $V(\cdot)$ would require its entire finite dimensional distribution (Cressie, 1991, p. 52), this section only defines its first and second moments, which are assumed to exist. A second order description provides sufficient information when $V(\cdot)$ is approximately Gaussian (Vanmarcke, 1983, p. 5). In the general situation higher order moments should also be considered, but this is much more complicated (Guardiano and Srivastava, 1993).

Let x and x' be elements of D. The correlation $\rho(x,x')$ of $V(x)$ and $V(x')$ is defined as:

$$\rho(x,x') = \frac{R(x,x')}{\sigma(x)\sigma(x')} \tag{2.2}$$

where $R(x,x')$ is the covariance of $V(x)$ and $V(x')$. Clearly, when $x = x'$ then covariance equals variance, so $R(x,x) = \sigma^2(x)$ and $\rho(x,x) = 1$ for all $x \in D$. In order for the autocovariance function $R(\cdot,\cdot)$ and autocorrelation function $\rho(\cdot,\cdot)$ to be valid functions, $R(\cdot,\cdot)$ must be symmetric and positive definite (Cressie, 1991).

Multivariate extension

Consider the multivariate extension of the error model (2.1). Now there are multiple attributes $A_i(x)$ and errors $V_i(x)$, $i = 1, \ldots, m$. For each of the attributes an error model $A_i(x) = b_i(x) + V_i(x)$ is defined, where the error $V_i(x)$ follows some distribution with mean $\xi_i(x)$ and variance $\sigma_i^2(x)$. Let $\rho_{ij}(x,x')$ be the correlation of $V_i(x)$ and $V_j(x')$, defined as:

$$\rho_{ij}(x,x') = \frac{R_{ij}(x,x')}{\sigma_i(x)\sigma_j(x')} \qquad (2.3)$$

where $R_{ij}(x,x')$ is the covariance of $V_i(x)$ and $V_j(x')$. The cross-covariance function $R_{ij}(\cdot,\cdot)$ thus defines the covariance of different attribute errors, possibly at different locations. The matrix of covariance functions $R_{ij}(\cdot,\cdot)$ must be positive definite (Journel and Huijbregts, 1978, p. 171). Figure 2.1 graphically illustrates the differences between the correlations $\rho_{ij}(x,x)$, $\rho_{ii}(x,x')$ and $\rho_{ij}(x,x')$.

Of particular interest is the correlation between attributes at the same location. From the definition of the error model it follows that the correlation $\rho_{ij}(x,x)$ of the errors $V_i(x)$ and $V_j(x)$ is identical to the correlation of the attributes $A_i(x)$ and $A_j(x)$. However, it is useful to note that the correlation of $A_i(x)$ and $A_j(x)$ should not be confused with the correlation between spatial attributes, as it is ordinarily computed from paired comparisons of attribute values at different locations in an area. The difference between the two will become clear in section 2.4.

To illustrate that errors are often correlated, consider the example of soil texture. If the texture classes sand, silt and clay are defined such that they are mutually exclusive and jointly exhaustive, then for all $x \in D$ these should satisfy the identity $A_{\text{sand}}(x) + A_{\text{silt}}(x) + A_{\text{clay}}(x) = 100$ per cent.

The maps of sand, silt and clay content will probably be constructed such that $b_{\text{sand}}(x) + b_{\text{silt}}(x) + b_{\text{clay}}(x) = 100$ for all $x \in D$. As a result we get $V_{\text{sand}}(x) + V_{\text{silt}}(x) + V_{\text{clay}}(x) = 0$. From this identity it follows that $\text{var}(V_{\text{sand}}(x) + V_{\text{silt}}(x) + V_{\text{clay}}(x)) = 0$. This implies that when we quantify the maps of variances and

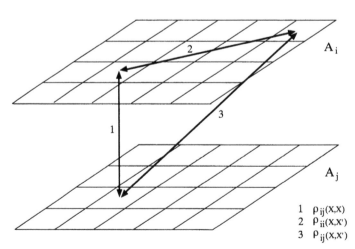

Figure 2.1 Graphical illustration of the differences between the correlations $\rho_{ij}(x, x)$, $\rho_{ii}(x, x')$ and $\rho_{ij}(x, x')$.

correlations we must guarantee that for all $x \in D$:

$$\sigma^2_{\text{sand}}(x) + \sigma^2_{\text{silt}}(x) + \sigma^2_{\text{clay}}(x) + 2\rho_{\text{sand-silt}}(x)\sigma_{\text{sand}}(x)\sigma_{\text{silt}}(x)$$

$$+ 2\rho_{\text{sand-clay}}(x)\sigma_{\text{sand}}(x)\sigma_{\text{clay}}(x) + 2\rho_{\text{silt-clay}}(x)\sigma_{\text{silt}}(x)\sigma_{\text{clay}}(x) = 0 \qquad (2.4)$$

To satisfy (2.4), it follows that negative error correlations must prevail. In fact, this result is quite conceivable, because when the error for sand is positive (when the real sand content is larger than the map value), then either silt or clay or both must make up for this, by taking on a lower value.

Another example of correlated errors is taken from soil pollution by heavy metals, such as is the case in the river Geul valley, in the south of the Netherlands (Leenaers, 1991). Consider the concentration of lead and cadmium in the soil, maps of which are obtained from interpolating point observations. In this case the interpolation errors $V_{\text{lead}}(x)$ and $V_{\text{cadmium}}(x)$ are likely to be positively correlated, because *unexpectedly* high lead concentrations will often be accompanied by *unexpectedly* high cadmium concentrations. Unforeseen low concentrations will also often occur simultaneously. Section 2.4 derives these error correlations mathematically for geostatistical interpolation.

The examples above thus illustrate that attribute errors are often correlated. This is an important observation because in later chapters we will see that the presence of correlation can have a marked influence on the outcome of an error propagation analysis.

2.3   Identification of the error model

To estimate the parameters of the error random field $V(\cdot)$ in practice, certain stationarity and ergodicity assumptions have to be made (Matheron, 1989, p. 81; Cressie, 1991, p. 53). This can be done in different ways. The most obvious way is to impose the assumptions directly on $V(\cdot)$. This is acceptable when inference on $V(\cdot)$ is based solely on observed errors at test points. For instance, to assess the error standard deviation of an existing DEM it may be sensible to assume that $\sigma(\cdot)$ is spatially invariant, so that it can be estimated by the root mean square error (RMSE), which is computed from the differences between the DEM and the true elevation at the test points (Fisher, 1992).

However, in many situations it is not very sensible to impose the stationarity assumptions directly on the error map $V(\cdot)$. In many situations $V(\cdot)$ is the residual from mapping an attribute whose *spatial variability* has been identified prior to, and has been incorporated in the mapping. In order to avoid inconsistencies, the error model parameters should then be derived from the spatial variability of the attribute and the mapping procedure used. This can be done in a unique manner if the spatial variability of the attribute and the mapping procedure are properly defined.

In this section I show how to identify the error model from the spatial variability of the attribute and the mapping procedure used. But first I must introduce and explain the idea of spatial variability, and how the variability of a spatial attribute relates to its uncertainty.

2.3.1 Three models of spatial variation

The general model of spatial variation (GMSV) considers a spatial attribute $Z(\cdot)$ given by:

$$Z(x) = \mu(x) + \varepsilon(x) \quad \text{for all } x \in D \tag{2.5}$$

where $\mu(\cdot)$ is the (deterministic) mean of $Z(\cdot)$, otherwise known as the trend, representing large-scale fluctuation, and $\varepsilon(\cdot)$ is a zero-mean, spatially autocorrelated residual, representing small-scale fluctuation (Cressie, 1991).

In general, additional assumptions on $\mu(\cdot)$ and $\varepsilon(\cdot)$ are needed to facilitate their identification from point observations. In this section I first discuss two simple versions of the GMSV, known as the *discrete model of spatial variation* (DMSV) and the *continuous model of spatial variation* (CMSV) (Bregt, 1992; Anselin *et al.*, 1994). I will also present a combination of the DMSV and CMSV, denoted here as the *mixed model of spatial variation* (MMSV) (Heuvelink, 1996; Heuvelink and Huisman, 1996).

The discrete model of spatial variation

The DMSV first divides the domain D into K mutually exclusive units D_k $(k = 1, \ldots, K)$. It then makes the following assumptions on the GMSV (2.5):

1 $\mu(x) = \beta(k)$ for all $x \in D_k$ $(k = 1, \ldots, K)$.

2 $\varepsilon(\cdot)$ is spatially uncorrelated.

3 $\text{var}(\varepsilon(x)) = C_0$ for all $x \in D$.

Thus the DMSV assumes that the area of interest D can be divided into a finite number of more or less homogeneous units D_k, and that the value of $Z(x)$ at any location x in unit D_k is the sum of a unit-dependent mean $\beta(k)$ and a residual noise term $\varepsilon(x)$, which is uncorrelated with neighbouring values and whose variance is constant throughout the domain D. The residual noise represents the within-unit variability (Burrough, 1986, p. 108; Voltz and Webster, 1990). It is worth noting that the DMSV has not been specifically developed for spatial attributes, but stems from the statistical model underlying analysis 'of variance methods (Burrough, 1986, p. 139; Snedecor and Cochran, 1989; Casella and Berger, 1990).

The DMSV will usually be adopted when the units D_k are available in the form of a polygon map, such as a soil map, a geological map or a vegetation map, and when the within-unit variability is expected to be small in comparison with the between-unit variability. In other words, the DMSV is appropriate when major jumps in the attribute $Z(\cdot)$ take place at the boundaries of the mapping units (Voltz and Webster, 1990; Burrough, 1993). For instance, many physical and chemical soil properties depend on soil type and so the DMSV may be adopted using the soil map to partition the area (Beckett and Burrough, 1971; Van Kuilenburg *et al.*, 1982; Wösten *et al.*, 1985; Bregt and Beemster, 1989; Yost *et al.*, 1993; Kern, 1994).

The continuous model of spatial variation

In its simplest form, the CMSV makes the following assumptions on the GMSV to facilitate its identification in practice:

1 $\mu(x) = \mu$ for all $x \in D$.

2 $\text{cov}(Z(x), Z(x + h)) = C_Z(h)$ for all $x, x + h \in D$.

Thus the CMSV assumes that $Z(\cdot)$ is a second order stationary random field, meaning that there is no trend, and that the autocovariance function of $Z(\cdot)$, denoted here by $C_Z(\cdot)$, is a function only of the distance between the locations (Oliver and Webster, 1990; Cressie, 1991). Note that the latter assumption does not exclude the presence of anisotropy, in which case the covariance of $Z(x)$ and $Z(x + h)$ depends on both the magnitude *and* the direction of the distance vector h (Journel and Huijbregts, 1978; Cressie, 1991).

Central to the CMSV is the concept of spatial autocorrelation. In the geostatistical literature it is customary to use the variogram $\gamma_Z(\cdot)$ to characterise the spatial autocovariance structure of $Z(\cdot)$:

$$\gamma_Z(h) = \tfrac{1}{2}E[(Z(x) - Z(x + h))^2] \quad \text{for all } x, x + h \in D \tag{2.6}$$

which has the advantage that it can be estimated from point observations of $Z(\cdot)$, without reference to μ. Characteristic parameters of the variogram are the nugget, sill and range (for a detailed discussion of variogram parameters, see e.g. Cressie, 1991). Note that when $Z(\cdot)$ is second order stationary, then the autocovariance function and the variogram are directly related by the identity $\gamma_Z(h) = C_Z(0) - C_Z(h)$. Here a notation in terms of the autocovariance function $C_Z(\cdot)$ is maintained.

The mixed model of spatial variation

The assumptions underlying the MMSV are a combination of those underlying the DMSV and CMSV:

1 $\mu(x) = \beta(k)$ for all $x \in D_k$.

2 $\text{cov}(\varepsilon(x), \varepsilon(x + h)) = C_\varepsilon(h)$ for all $x, x + h \in D$.

Second order stationarity is thus imposed on $\varepsilon(\cdot)$ instead of on $Z(\cdot)$. The MMSV is more general than the DMSV and CMSV, and in fact it contains both models. However, the DMSV and CMSV are included here as separate models because they are very often used in practice. In section 2.3.2 we will see that identification of the error model under the MMSV is somewhat more difficult, which may be the reason why this model has not yet been frequently used.

Comparison of the three models of spatial variation

A comparison of the assumptions underlying the three models of spatial variation shows that the principal difference is in the way they approach spatial variation. Whereas the DMSV assumes that major jumps in attribute values occur at the boundaries of homogeneous mapping units, the CMSV assumes that attribute values change gradually in space. The MMSV is a mixture of both and includes gradual as well as abrupt changes. The choice of which model to employ in a given situation should thus largely depend on the kind of spatial variability at hand. This is confirmed by experience from soil science, where it generally is found that the continuous model performs poorly where there are sharp boundaries between soil units, and where the discrete model is lacking when soil variation is gradual (Voltz and Webster, 1990).

The difference between the three models of spatial variation is graphically illustrated in figure 2.2.

2.3.2   Error identification under the three models of spatial variation

We now turn to the situation in which a spatial attribute $Z(\cdot)$ is to be mapped from point observations $z(x_i)$, $i = 1, \ldots, n$. The result of the mapping will be different depending on which model of spatial variation is used, but the principle is the same. In statistical terms, the mapping is done by *conditioning* the random field $Z(\cdot)$ to the observations $z(x_i)$, yielding a prediction $z^*(x)$ and a prediction error $Z(x) - Z^*(x)$ for all $x \in D$:

$$\{Z(x) \,|\, Z(x_i) = z(x_i), i = 1, \ldots, n\} = \{Z^*(x) \,|\, Z(x_i) = z(x_i), i = 1, \ldots, n\}$$

$$+ \{Z(x) - Z^*(x) \,|\, Z(x_i) = z(x_i), i = 1, \ldots, n\}$$

$$= z^*(x) + (Z(x) - Z^*(x)) \tag{2.7}$$

Strictly speaking, the last equality in (2.7) is only valid when the predictor $Z^*(x)$ is completely identified by the observations and when the prediction error $Z(x) - Z^*(x)$ is statistically independent of the $Z(x_i)$. In fact, these are

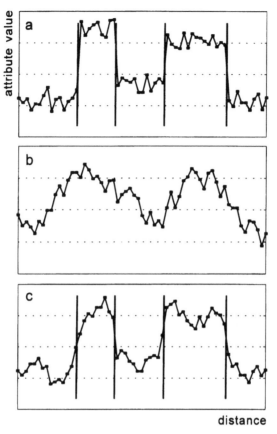

Figure 2.2 Example realisations along a transect of a spatial attribute satisfying the: (a) DMSV, (b) CMSV, (c) MMSV. Vertical lines represent boundaries between mapping units.

exactly the conditions that one will have in mind when separating $Z(x)$ into the two terms $Z^*(x)$ and $Z(x) - Z^*(x)$, but the separation may be hampered by linearity and unbiasedness conditions (Cressie, 1991, p. 177).

In the notation used in section 2.2, the conditional random variable $\{Z(x) \mid Z(x_i) = z(x_i),\ i = 1, \dots, n\}$ equals the error-contaminated attribute $A(x)$, which is the sum of the prediction $b(x) = z^*(x)$ and the prediction error $V(x) = Z(x) - Z^*(x)$.

It is important to underline that although $Z(\cdot)$ and $A(\cdot)$ describe the *same* spatial attribute, there is a fundamental difference between the two. Whereas the stochasticity of $Z(\cdot)$ refers to the *spatial variability* of the attribute, the stochasticity of $A(\cdot)$ represents the *uncertainty* about it (Delhomme, 1979). The uncertainty is not only clearly affected by the spatial variability, but also by the density of observations and the mapping procedure used. For instance, an increase in the number of observations will not affect the spatial variability of the attribute, but it will reduce the uncertainty about it.

Case of the DMSV

When $Z(\cdot)$ satisfies the DMSV and when $\beta(\cdot)$ is unknown, it is sensible to use the within-unit sample mean as a predictor of the value of $Z(x)$:

$$Z_D^*(x) = \frac{1}{n_k} \sum_{i=1}^{n_k} Z(x_i) \quad x, x_i \in D_k \tag{2.8}$$

where n_k is the number of observations located in unit D_k. Thus the actual observations $z(x_i)$ are substituted into (2.8) for all $x \in D_k$ to construct the map $b(\cdot)$. The mapping error $V(x) = Z(x) - Z_D^*(x)$ will have zero mean and variance:

$$\sigma_D^2(x) = \text{var}(Z(x) - Z_D^*(x)) = \left(1 + \frac{1}{n_k}\right)C_0 \quad x \in D_k \tag{2.9}$$

In practice the variance C_0 is often unknown, so it will be estimated by weighted averaging of the per-unit sample variances. Estimating $\beta(k)$ from the sample mean also has an effect on the spatial autocorrelation of the error $V(\cdot)$. The autocorrelation is no longer zero for points lying in the same unit. Instead, we get $\rho(x,x') = 1/(n_k + 1)$ for $x, x' \in D_k, x \neq x'$.

In formulating the DMSV we have assumed that the within-unit variance C_0 is the same for all units. Alternatively, this assumption may be weakened by allowing the variance to vary between units. The advantage is a more general model, but the disadvantage is that per-unit variances are based on smaller samples and are thus less accurate. The assumption of equal variances made here makes it easier to compare the DMSV with the CMSV and MMSV.

Case of the CMSV

An attribute that satisfies the CMSV will be mapped using geostatistical interpolation, otherwise known as kriging (Oliver and Webster, 1990; Cressie, 1991). Restricting ourselves to ordinary kriging, the predictor $Z_C^*(x)$ of $Z(x)$ will be taken as a weighted linear combination of the observations $Z(x_i)$:

$$Z_C^*(x) = \sum_{i=1}^{n} \lambda_i Z(x_i) \quad x, x_i \in D \tag{2.10}$$

where the weights λ_i are chosen such that the variance of the prediction error $Z(x) - Z_C^*(x)$ is minimised, under the condition of unbiasedness ($\sum \lambda_i = 1$). The weights are determined from the autocovariance function $C_Z(\cdot)$ and the configuration of the locations x_i and x. Note that the autocovariance $C_Z(\cdot)$ or variogram $\gamma_Z(\cdot)$ is assumed to be known. It can be estimated from the experimental variogram that is computed from the observations. The kriging prediction error variance is given by:

$$\sigma_C^2(x) = \text{var}(Z(x) - Z_C^*(x))$$

$$= C_Z(0) + \sum_{i=1}^{n} \sum_{j=1}^{n} \lambda_i \lambda_j C_Z(x_i - x_j) - 2 \sum_{i=1}^{n} \lambda_i C_Z(x_i - x)$$

$$x, x_i \in D \qquad (2.11)$$

Thus in this case the map $b(\cdot) = z_C^*(\cdot)$ is computed from substituting the observations $z(x_i)$ in (2.10), whereas the variance of the error map $V(\cdot)$ is derived from (2.11). The kriging variance $\sigma_C^2(x)$ is usually much smaller than the *a priori* variance $C_Z(0)$ of $Z(x)$, which demonstrates that by conditioning the attribute to the observations its uncertainty can be reduced to a level that is below its spatial variability.

Let us now consider the autocorrelation $\rho(\cdot,\cdot)$ of the mapping error $V(\cdot)$. Recall from (2.2) that $\rho(x,x') = R(x,x')/(\sigma(x)\sigma(x'))$, where in this case the standard deviations $\sigma(x)$ and $\sigma(x')$ are the kriging standard deviations $\sigma_C(x)$ and $\sigma_C(x')$. The covariance of the errors $V(x)$ and $V(x')$ is given by:

$$R(x,x') = E[(Z(x) - Z_C^*(x))(Z(x') - Z_C^*(x'))] \qquad (2.12)$$

which can straightforwardly be computed from the autocovariance function (or variogram) and the kriging weights, yielding (Delhomme, 1979):

$$R(x,x') = C_Z(x - x') + \sum_{i=1}^{n} \sum_{j=1}^{n} \lambda_i v_j C_Z(x_i - x_j)$$

$$- \sum_{i=1}^{n} \lambda_i C_Z(x_i - x') - \sum_{j=1}^{n} v_j C_Z(x_j - x) \qquad (2.13)$$

Here the v_j are the kriging weights of $Z_C^*(x')$. A problem now is that the so-obtained autocorrelation $\rho(\cdot,\cdot)$ is unlikely to vary with only the distance $h = x - x'$. To store it would therefore require storage of $\rho(x,x')$ for all combinations of x and x', and this is hardly practically feasible. A better solution might be to store the autocovariance function $C_Z(\cdot)$ of $Z(\cdot)$ and the kriging weights for all locations, so that $R(x,x')$ and $\rho(x,x')$ can be recomputed whenever necessary.

Case of the MMSV

The mapping of a spatial attribute that satisfies the MMSV is done using universal kriging (Christensen, 1991; Cressie, 1991), sometimes also referred to as kriging with an external drift (Deutsch and Journel, 1992, p. 67). One of the main problems with universal kriging has always been the assessment of the variogram $\gamma_\varepsilon(\cdot)$ or autocovariance function $C_\varepsilon(\cdot)$ of $\varepsilon(\cdot)$. The problem is that estimation of $\gamma_\varepsilon(\cdot)$ requires knowledge of $\beta(\cdot)$, but efficient estimation of $\beta(\cdot)$ in turn requires knowledge of $\gamma_\varepsilon(\cdot)$ (Cressie, 1991, p. 165). Direct estimation of $\gamma_\varepsilon(\cdot)$ from detrended data has been criticised in the past because it yields a distorted variogram (Armstrong, 1984), and complicated iterative methods were proposed instead. However, in a recent paper it was shown that direct estimation of $\gamma_\varepsilon(\cdot)$ from detrended data is viable, because the distorted variogram does not impair the outcome of the kriging procedure (Kitanidis, 1994).

Application of universal kriging yields predictions and prediction error variances for all $x \in D$. The difference with ordinary kriging is that the presence of a trend is accounted for. Thus universal kriging simultaneously estimates the coefficients $\beta(k)$ and predicts the residual $\varepsilon(\cdot)$. Now since both the estimator of the trend and the predictor of the residual are linear in the observations, in effect the prediction of $Z(x)$ will again be a linear combination of the observations:

$$Z_M^*(x) = \sum_{i=1}^{n} \kappa_i Z(x_i) \quad x, x_i \in D \tag{2.14}$$

The universal kriging weights κ_i will differ from the ordinary kriging weights λ_i for two main reasons. Firstly, the universal kriging equations that are solved to obtain the weights are based on the variogram $\gamma_\varepsilon(\cdot)$ instead of on $\gamma_Z(\cdot)$. Secondly, additional unbiasedness conditions are imposed to ensure that the κ_i sum to one for all points lying in the same unit as x, and that they sum to zero for all other units individually.

The universal kriging variance is analogous to (2.11):

$$\sigma_M^2(x) = C_\varepsilon(0) + \sum_{i=1}^{n} \sum_{j=1}^{n} \kappa_i \kappa_j C_\varepsilon(x_i - x_j) - 2 \sum_{i=1}^{n} \kappa_i C_\varepsilon(x_i - x) \quad x, x_i \in D \tag{2.15}$$

The spatial autocorrelation of the universal kriging prediction error is computed in a similar way as with ordinary kriging.

2.4 Multivariate extension

The multivariate extension of the error identification is important here because it provides a means to quantify the cross-correlation function $\rho_{ij}(\cdot, \cdot)$ of the errors $V_i(\cdot)$ and $V_j(\cdot)$. In chapters 4 and 5 it will become clear that knowledge about the cross-correlation is imperative in order to adequately carry out an error propagation analysis involving multiple attributes.

The discussion on the multivariate extension of the error identification will be restricted here to the CMSV, but a generalisation to the DMSV and MMSV can be made without major difficulties.

Cokriging in matrix notation

The multivariate CMSV extends kriging to cokriging, which is a form of kriging that allows the inclusion of multiple attributes in the prediction. Consider m second order stationary random fields $Z_j(\cdot)$, $j = 1, \ldots, m$, that are sampled at n locations $x_i \in D$, $i = 1, \ldots, n$. Let $\bar{Z}(x)$ be the row vector $[Z_1(x) \cdots Z_m(x)]$ that is to be predicted from the observations $\bar{Z}(x_1), \ldots, \bar{Z}(x_n)$. Let $\bar{C}(\cdot)$ be the $m \times m$ matrix of covariance functions $C_{ij}(\cdot)$, where:

$$C_{ij}(h) = \text{cov}(Z_i(x), Z_j(x + h)) \quad x, x + h \in D \tag{2.16}$$

Then the best linear unbiased cokriging predictor $\bar{Z}^*(x)$ of $\bar{Z}(x)$ is given by (Myers, 1982):

$$\bar{Z}^*(x) = \sum_{i=1}^{n} \bar{Z}(x_i)\Lambda_i \tag{2.17}$$

where each Λ_i is an $m \times m$ matrix of cokriging weights that are the solution of the linear system:

$$
\begin{bmatrix}
\bar{C}(x_1 - x_1) & \cdots & \bar{C}(x_1 - x_n) & I \\
\vdots & \vdots & \vdots & \vdots \\
\bar{C}(x_n - x_1) & \cdots & \bar{C}(x_n - x_n) & I \\
I & \cdots & I & 0
\end{bmatrix}
\cdot
\begin{bmatrix}
\Lambda_1 \\
\vdots \\
\Lambda_n \\
L
\end{bmatrix}
=
\begin{bmatrix}
\bar{C}(x - x_1) \\
\vdots \\
\bar{C}(x - x_n) \\
I
\end{bmatrix}
\tag{2.18}
$$

with L an $m \times m$ matrix of Lagrange multipliers, I the $m \times m$ identity matrix, 0 the $m \times m$ null matrix.

The covariance matrix $\bar{R}(x,x)$ of the prediction error $(\bar{Z}(x) - \bar{Z}^*(x))$ is given by:

$$\bar{R}(x,x) = \text{var}(\bar{Z}(x) - \bar{Z}^*(x)) = \bar{C}(0) - \sum_{i=1}^{n} \bar{C}(x - x_i)\Lambda_i - L \tag{2.19}$$

The cokriging predictions and prediction error variances for each of the $Z_i(\cdot)$ are thus obtained from (2.17) and (2.19), after first having solved the cokriging system (2.18). The correlations of the cokriging prediction errors at x are computed from $\bar{R}(x, x)$ as:

$$\rho_{ij}(x,x) = \frac{\bar{R}(x,x)_{ij}}{\sqrt{(\bar{R}(x,x)_{ii}\,\bar{R}(x,x)_{jj})}} \tag{2.20}$$

The correlations of the cokriging prediction errors will generally vary spatially, but an exception is when the random functions all have the same spatial correlation structure, i.e. when the random functions are in *intrinsic coregionalisation* (Journel and Huijbregts, 1978, p. 174). In that case the covariance functions $C_{ij}(\cdot)$ are all equal up to a multiplicative constant:

$$C_{ij}(\cdot) = c_{ij}\,C_d(\cdot) \tag{2.21}$$

where $C_d(\cdot)$ is a 'direct' covariance function. It is not difficult to show that this causes the correlations of the cokriging errors to be greatly simplified to:

$$\rho_{ij}(x,x) = \frac{c_{ij}}{\sqrt{(c_{ii}c_{jj})}} \tag{2.22}$$

which are spatially invariant and in addition equal to the correlations of the $Z_i(\cdot)$ and $Z_j(\cdot)$. However, note that this is true only when the $Z_i(\cdot)$ are in intrinsic coregionalisation *and* when no attribute is undersampled (Journel and Huijbregts, 1978, p. 175). Sections 5.3 and 5.4 give examples where the correlations of the cokriging prediction errors are not spatially invariant.

The correlation $\rho_{ij}(x,x')$ of the cokriging prediction errors $Z_i(x) - Z_i^*(x)$ and $Z_j(x') - Z_j^*(x')$ can be computed in much the same way as $\rho_{ij}(x,x)$. The resulting $\rho_{ij}(x,x')$ will generally vary with both x and x', so that it cannot be generalised to a function of only the distance $h = x - x'$. This causes similar storage problems to those discussed before for $\rho(x,x')$ in section 2.3.2.

2.5   Change of support issues

Up to now, no mention has been made of the support of the observations $z(x_i)$. However, the statistical properties of the spatial attribute $Z(\cdot)$ strongly depend on the size, shape and orientation of the observations (Journel and Huijbregts, 1978; Starks, 1986; Webster and Oliver, 1990; Heuvelink, 1998), which is known as the *support*. Synonyms for support that are sometimes used are aggregation level and observation scale. The prediction methods described in sections 2.3 and 2.4 implicitly assumed that all observations are collected on the same support. Moreover, the resulting error model parameters apply *only* to the same support.

Therefore, when the objective is to map the attribute at a larger support, then the change of support will have to be accounted for. This is a situation that is frequently observed in practice. For instance, in distributed hydrological modelling, grid cell averages of soil properties are required that are much larger than the 'point' values that are measured (Klemes, 1986; Beven, 1989; De Roo et al., 1992; Grayson et al., 1992b, 1993). In mining and soil pollution, measurements are usually obtained at supports that are much smaller than the volumetric units that are to be extracted (Isaaks and Srivastava, 1989, p. 461; Okx et al., 1990).

It is important to note that a change of support is understood here as computing the *arithmetical* mean of the attribute over an area or volume that is larger than the measurement support. The problem becomes much more complex when the aim is to compute nonlinear averages, such as is the case with so-called 'effective' parameters (Cressie, 1993; Gómez-Hernández, 1993).

Ignoring support problems need not be very harmful as long as one is interested in the kriged map $b(\cdot)$, but a change of support often has a marked influence on the error map $V(\cdot)$ (Isaaks and Srivastava, 1989; Webster and Oliver, 1990, p. 271). Consequently, because this research is primarily concerned with the uncertainty in spatial attributes, it is essential to pay attention to this particular problem.

Variance reduction with increase of support size

In general, the error variance will decrease due to a change of support to a larger volume or 'block' (Cressie, 1991, p. 284). This is because the spatial variation within the larger support is cancelled out, leaving less uncertainty about its average value.

An extreme case of variance reduction occurs with the discrete model of spatial variation. If the point support is much smaller than the block support, then the only uncertainty that effectively remains is the error in the mean. This means that the error variance (2.9) is reduced to C_0/n_k. However, one should question whether this result has any real meaning. It rather appears that it brings to bear an important shortcoming of the DMSV, namely the assumption of complete absence of spatial correlation.

With the continuous and mixed models of spatial variation the variance reduction due to an increase of support size is generally less dramatic, but it may still be substantial, particularly when there is a large nugget. Kriging to a larger support is known as block kriging (Isaaks and Srivastava, 1989; Webster and Oliver, 1990). Block kriging variances are thus smaller than point kriging variances, which appears attractive because it implies a reduction of uncertainty. However, one should bear in mind that the reduction of uncertainty is obtained for a 'different' attribute, which may or may not be intrinsically of interest (Webster and Oliver, 1990, p. 271).

Although the theory on change of support allows one to obtain the error model parameters for larger supports, departures from the model assumptions may render the results of such an approach unreliable. The DMSV is a clear example of this, but the CMSV and MMSV suffer from the same problem. The block kriging variance strongly depends on the behaviour of the variogram near the origin, and this is an interval where the variogram is often poorly estimated (Webster and Oliver, 1990, p. 272). It is much safer to use sample data that have the same support as the blocks to be estimated (Isaaks and Srivastava, 1989, p. 462). Unfortunately, the latter may not always be practically feasible (Finke *et al.*, 1998).

The design-based approach for very large support

When the support is so large that multiple observations are located within each block or area, then areal means can also be estimated in an entirely different way, using well-known methods from classical sampling theory (Cochran, 1977). For instance, consider the situation in which the objective is to estimate map-unit means of a polygon map. If multiple observations of the attribute are contained in each map unit, then a *design-based* approach yields estimates of the unit means, together with a measure of accuracy, provided the observations are collected using probability sampling (De Gruijter and Ter Braak, 1990; Ten Cate *et al.*, 1997). Design-based strategies can also be used to estimate the overall accuracy of maps generated from remotely sensed data (Congalton, 1988; Stehman, 1992).

The design-based approach has an important advantage over the *model-based* approach that I used throughout this chapter. This is that it makes very few assumptions about the spatial attribute under study. The attribute is merely seen as an unknown deterministic function, whereas the model-based approach imposes a non-existent stochastic 'superpopulation' model (Cressie, 1991). Thus the design-based approach is automatically protected against unrealistic model assumptions, and so it will often be favoured when the objective is to estimate spatial means (Papritz and Webster, 1995; Ten Cate *et al.*, 1997). However, its usefulness is rather limited when it comes to predicting values at points (Domburg *et al.*, 1993; Ten Cate *et al.*, 1997), which is the main reason why the design-based approach was not used in this research.

CHAPTER THREE

Identification of the error model: a case study

In this chapter the theory of the previous chapter is applied to a case study, in which the mean highest water table (MHW) in a polder area near Nijmegen, The Netherlands, is mapped from point observations. Clearly, from an error analysis perspective, the interest is not only in the interpolated map of the MHW itself, but also in the associated error map.

An important problem that will also be addressed in this chapter is how to decide which of the three models of spatial variation to adopt in a practical situation such as this, when no prior information about the type of spatial variation to be expected is available. In the case study we will use all three models of spatial variation to demonstrate how the choice between the models may be based on a validation procedure.

3.1 Mapping the mean highest water table in the Ooypolder

The area considered is the 'Ooypolder', which is approximately 27 km^2 in size and situated in the east of The Netherlands. The majority of soil profiles found here are entisols (Soil Survey Staff, 1975). In The Netherlands these soil profile classes are further subdivided into relatively dry and relatively wet entisol types, depending on their parent material and their elevation above the water table. These soil types are contained in a general 1 : 50000 soil map of the study area (Netherlands Soil Survey Institute, 1975).

Two important attributes of the 1 : 50000 soil map are the mean highest water table (MHW) and mean lowest water table (MLW) (Van Der Sluijs and De Gruijter, 1985). The MHW (cm) and MLW (cm) are calculated from long-term records of water tables. The MHW and MLW are important for describing the soil moisture condition, the soil aeration and the potential depth of the root zone. In the 1 : 50000 soil map, the MHW and MLW are mapped as 'water table classes', numbered I to VII. Both MHW and MLW tend to be larger in the upper classes. Figure 3.1 shows a map of the study area with the delineations of the water table classes. Hereafter attention will be focused on the MHW only.

In 1987 a survey was conducted in the study area. This survey was required for a specific land use sanitation project. Maps of soil properties were needed for this purpose that were far more accurate than the 1 : 50000 soil map. This survey resulted, among others, in a total number of 1987 observations of the MHW, regularly spread over the study area. In the analysis hereafter we use subsets of sizes 200 and 500 to map the MHW and its error using the DMSV, CMSV and MMSV. The performance of the methods will then be evaluated using a validation set of 993 observations, randomly selected from the remaining observations.

For each water table class the sample mean and variance of the MHW were computed from the subsets of 200 and 500 data points. The results are given in table 3.1. From the sample variances the general variance C_0 was estimated as 451 cm^2 for subset 200 and 450 cm^2 for subset 500. Notice that the figures in table 3.1 suggest that the variance is not the same for all water table classes, but notice also that the assumption of a per-unit variance would yield unreliable estimates for the lower classes, due to the small sample sizes.

Mapping the MHW using the DMSV, CMSV and MMSV

Next the DMSV was adopted to compute predictions and prediction error standard deviations of the MHW from the 500 subset. The resulting maps are given in figures 3.3a and 3.4a. Similar maps are obtained when the 200 subset is used. The map of standard deviations in figure 3.4a shows little spatial dif-

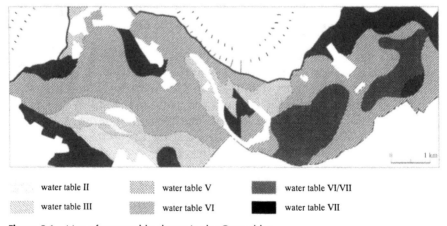

water table II	water table V	water table VI/VII
water table III	water table VI	water table VII

Figure 3.1 Map of water table classes in the Ooypolder.

Table 3.1   Sample size, sample mean (cm) and sample variance (cm^2) of MHW per water table class computed from the subsets of sizes 200 and 500.

Water table	Subset 200			Subset 500		
	Size	Mean	Variance	Size	Mean	Variance
II	3	10.0	100	3	10.0	100
III	2	22.5	113	9	18.3	81
V	34	43.1	646	81	35.7	503
VI	101	54.2	444	261	55.0	467
VI/VII	31	71.3	307	77	68.4	277
VII	29	57.6	440	69	63.7	565

ferentiation. This is as expected, because the only cause for spatial differentiation lies in the sample sizes used to estimate the per-unit mean (see (2.9)).

To perform ordinary and universal kriging from the 200 and 500 subsets, as required under the CMSV and MMSV respectively, structural analyses were carried out on the original and detrended data. The resulting variograms are given in figure 3.2. Parameters of the theoretical variograms that were fitted through the experimental variograms are given in table 3.2. All variograms are

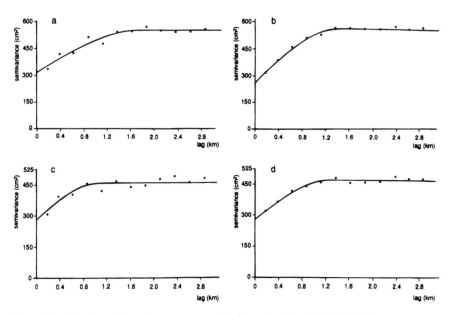

Figure 3.2   Experimental variograms computed from the: (a) original MHW observations, subset 200, (b) original MHW observations, subset 500, (c) detrended MHW observations, subset 200, (d) detrended MHW observations, subset 500. Solid lines represent the fitted variograms.

Table 3.2 Variogram parameters for the original and detrended MHW for subsets 200 and 500.

	Original data		Detrended data	
	Subset 200	Subset 500	Subset 200	Subset 500
Nugget (cm^2)	315	260	280	280
Sill (cm^2)	548	586	459	474
Range (km)	1.76	1.44	1.08	1.31

of the spherical type. The large nugget in the variograms is probably caused by the fact that measurement errors can be quite large. Notice that detrending has no obvious effect on the nugget but does cause a decrease of the sill and range. This is not surprising because subtracting the map-unit means from the

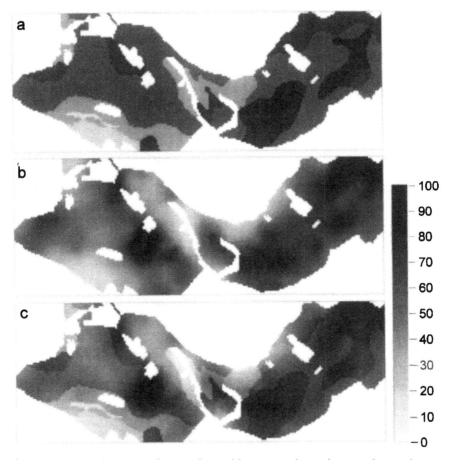

Figure 3.3 Maps of MHW predictions obtained from interpolating the 500 subset and using the: (a) DMSV, (b) CMSV, (c) MMSV.

MHW observations filters out global effects but does not markedly affect short-distance variation.

Maps of the ordinary and universal kriging predictions and prediction error standard deviations are given in figures 3.3b,c and 3.4b,c, again using the 500 subset. The kriging was done using the software package gstat (Pebesma and Wesseling, 1998), but alternatively the GSLIB package (Deutsch and Journel, 1992) could have been used. The maps of standard deviations show that kriging errors are large when near the border of the study area.

3.2   Comparison of mapping results for the three models of spatial variation

There are clear differences between the maps of MHW predictions in figure 3.3. Particularly the map with the DMSV predictions differs importantly from the others. This map necessarily follows the boundaries between the mapping units. However, closer inspection of the map with MMSV predictions shows that some of the boundaries between water table classes are recognisable in it. For instance, water table class II shows up in the south-west in figure 3.3c. Nonetheless, it is evident that the map of MMSV predictions much more agrees with its CMSV counterpart than with its DMSV counterpart. Both the MMSV and CMSV maps show a continuous behaviour, although the CMSV map still is somewhat smoother. Notice also that none of the three maps contains on average larger or smaller predictions.

Comparison of the error maps in figure 3.4 also shows clear differences in spatial pattern. Again the DMSV map is the one that differs importantly from the others. Most striking is that figure 3.4a contains on average much larger values. This suggests that the DMSV is the least suitable model for describing the spatial variation of the MHW. There is again little difference between the CMSV and MMSV maps, except for the influence of water table class II. The MMSV error is much larger in water table class II due to the small sample size used to estimate the mean of class II.

Validation results

In order to judge which model of spatial variation is more appropriate here, a validation was carried out using the validation set of 993 observations. The results are given in table 3.3. These show that none of the methods appear to be biased, which is in agreement with the safeguards that are contained in each of the mapping procedures. The SRMSE values are close to 1, which means that all three methods on average yield realistic error estimates. Thus the reported MSE and SRMSE results are similar for all three models. But the RMSE values clearly show that the CMSV and MMSV outperform the

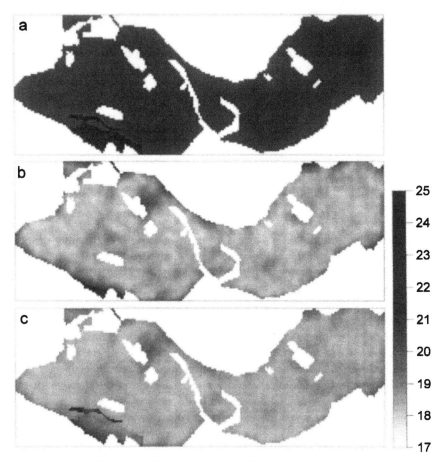

Figure 3.4 Maps of MHW prediction error standard deviations obtained from interpolating the 500 subset and using the: (a) DMSV, (b) CMSV, (c) MMSV.

Table 3.3 Validation results for mapping the MHW from the subsets of sizes 200 and 500.

	Mean error (cm)	Root mean square error (cm)	Standardised root mean square error
DMSV subset 200	0.27	22.9	1.07
CMSV subset 200	−0.34	21.1	1.04
MMSV subset 200	−0.26	20.8	1.05
DMSV subset 500	0.77	22.4	1.05
CMSV subset 500	0.62	19.1	1.02
MMSV subset 500	0.72	19.1	1.02

DMSV. To appreciate fully the differences between the RMSE values one must realise that the presence of a large nugget means that no method can yield an RMSE smaller than 16 cm.

The validation results confirm the observations made when the maps of standard deviations were discussed. Both the large RMSE values and the large standard deviations in figure 3.4a mean that the DMSV is not a very suitable model in this case. The CMSV and MMSV are more capable of modelling the spatial variation of the MHW. Comparison shows that there is little difference between the performance of the CMSV and MMSV. Apparently the MHW behaves predominantely as continuous, leaving only a marginal improvement by allowing discrete spatial changes as well. Given the fact that continuous spatial variation clearly dominates over discrete spatial variation, it is not surprising that the MMSV mainly relies on the CMSV.

Another interesting observation is that the CMSV and MMSV improve considerably when the number of observations is increased from 200 to 500. The DMSV fails to benefit from this, simply because the model is too rigid and can only use the extra information for a (marginal) improvement of the estimates of the unit means.

3.3 Discussion and implications for error propagation analysis

The case study clearly shows that when an attribute is to be mapped from point observations, then the underlying model of spatial variation has a clear influence on the resulting map and its error. This is important also from an error propagation analysis perspective. For instance, the results of an error analysis may be quite different depending on whether maps 3.3a and 3.4a or maps 3.3c and 3.4c are used as input to the analysis. Although not addressed in the case study, the spatial autocorrelation of error is also very much dependent upon the adopted model of spatial variation.

In practice the choice of a model of spatial variation will often be based on the information at hand: the DMSV is adopted when information is available in the form of a polygon map, the CMSV when all that is available are point observations (Heuvelink and Bierkens, 1992). The MMSV needs both types of information. Lack of data may thus force users to adopt a model that may well be inferior to others.

The choice of model may also be influenced by other factors, such as background and experience of the user. A traditional soil surveyor is brought up with the idea that a soil map should be built up of discrete mapping units. Likewise, a forest scientist may by habit delineate a forest area into forest stands, even when variability within stands is of the same order of magnitude as variability between stands. On the other hand, a geostatistician may be reluctant to include 'soft' information in the form of map delineations in a spatial interpolation procedure.

Another important obstacle against a free choice of model is that object-oriented vector data structures are often not very suited to deal with continuous spatial variation. This inclines users to opt for the DMSV. Raster GISs are more flexible in this respect and can easily handle all three types of spatial variation.

When compared with the DMSV and CMSV, it is obvious that the MMSV is attractive for modelling spatial attributes that show both discrete and continuous spatial behaviour. But the results also suggest that the MMSV should also be recommended for situations where there is no prior information about the kind of spatial variation present. This is because the MMSV is much more flexible than the DMSV and CMSV. This anticipated flexibility of the MMSV has indeed been confirmed by a simulation study (Heuvelink and Huisman, 1996).

Clearly the DMSV, CMSV and MMSV are not the only models available for modelling spatial variation. These models were addressed here because the DMSV and CMSV are most often used in practice and because the MMSV is a natural combination of the two that has many attractive properties. More elaborate models of spatial variation may be more appropriate in particular situations. It should be noted, however, that more complex models contain more parameters that somehow have to be estimated from the data. Admittedly there always remains some subjectivity in the choice of a model of spatial variation (Matheron, 1989). However, given that the choice of model has such a profound influence on the resulting map and its error, one would like to avoid subjective decisions as much as possible. Perhaps a test such as proposed by Kitanidis (1997) may assist in reaching the 'optimal' model in a more objective fashion.

CHAPTER FOUR

Error propagation with local GIS operations: theory

The purpose of this chapter is to trace the propagation of errors for one class of GIS operations, which have been identified in the first chapter by the name of *local* operations. This term refers to operations, for which the value of the output map at some location only depends on the value of the input maps at that same location, or possibly on their values within not too large a window surrounding that location. Local operations thus include point operations and neighbourhood operations (figure 4.1). Examples of local operations are Multiply, Maximise, Smooth, Sqrt and Slope (Burrough, 1986; Tomlin, 1990; Burrough and McDonnell, 1998), or a combination of these, such as a multiple regression model (Heuvelink *et al.*, 1989; Vereecken *et al.*, 1989), a fuzzy classification model (Burrough, 1989a; Burrough *et al.*, 1992) or a crop growth model (Van Diepen *et al.*, 1989; Van Lanen *et al.*, 1992).

Contrary to local operations, with *global* operations there can be far-reaching spatial interactions, so that the value of the output map at some

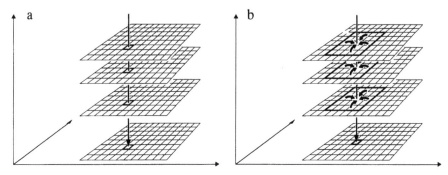

Figure 4.1 A local GIS operation can be a: (a) point operation or (b) neighbourhood operation.

location may be affected by input map values at remote locations. Examples of global operations are Spread, Stream and View (Burrough, 1986; Burrough and McDonnell, 1998). Distributed dynamic models, such as used to model groundwater flow (McDonald and Harbaugh, 1984), catchment hydrology (Abbott *et al.*, 1986; Grayson *et al.*, 1992a; Binley *et al.*, 1997) or erosion (Beasley and Huggins, 1980; De Roo, 1993) also include global interactions. Global operations require a somewhat different approach to analyse the propagation of errors, which is described in chapter 6.

Mathematical formulation of the error propagation problem

The error propagation problem can now be formulated as follows. Let the input maps of a GIS operation be random fields $A_i(\cdot)$ $(i = 1, \ldots, m)$, that are defined on some domain D, and that satisfy the equation $A_i(\cdot) = b_i(\cdot) + V_i(\cdot)$. Here $b_i(\cdot)$ is the map stored in the GIS and $V_i(\cdot)$ is a random field representing error or uncertainty. The output map $U(\cdot)$ is computed from the input maps $A_i(\cdot)$ by means of the GIS operation $g(\cdot)$:

$$U(\cdot) = g(A_1(\cdot), \ldots, A_m(\cdot)) \tag{4.1}$$

The output map $U(\cdot)$ is also a random field, with mean $\zeta(\cdot)$ and variance $\tau^2(\cdot)$. These are again functions defined on the domain D. In standard GIS analysis without error propagation the objective is to derive $\zeta(\cdot)$, which is estimated by $g(b_1(\cdot), \ldots, b_m(\cdot))$. However, when studying error propagation, the main interest goes to the uncertainty of $U(\cdot)$, as contained in its variance $\tau^2(\cdot)$. Sometimes knowledge about only the variance of $U(\cdot)$ is not sufficient, and other parameters of the distribution $F_{U(\cdot)}(\cdot)$ may be required as well.

In chapter 2 it was shown that the errors $V_i(\cdot)$ usually have zero mean, because unbiasedness conditions are included in the mapping procedure. From now on I will therefore assume that $\xi_i(x) = 0$ for all $x \in D$. In fact, this is quite a reasonable assumption, because if there were a *known* non-zero bias $\xi_i(\cdot)$ then it could easily be eliminated by adding it to $b_i(\cdot)$. If bias is *unknown*, then the only viable option seems to be to include it as an unknown zero-mean random component, which would merely increase the variance of $V_i(\cdot)$ (Schweppe, 1973).

It is worthwhile to note that when $g(\cdot)$ is nonlinear, the mean $\zeta(\cdot)$ of $U(\cdot)$ usually differs from $g(b_1(\cdot), \ldots, b_m(\cdot))$, even when the errors $V_i(\cdot)$ all have zero mean. This means that random errors in the input can cause systematic errors in the output. In fact, a marked example of this will be given in section 5.4.

4.1   Error propagation with point operations

The general problem stated above is relatively easy when $g(\cdot)$ is a point operation. With point operations there are no spatial interactions, which means

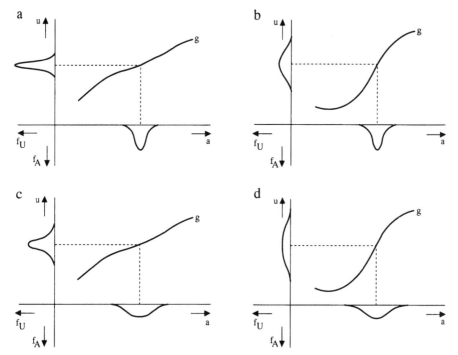

Figure 4.2 Functions of a random variable: (a) moderate $g(\cdot)$ with small variance σ^2, (b) steep $g(\cdot)$ with small variance σ^2, (c) moderate $g(\cdot)$ with large variance σ^2, (d) steep $g(\cdot)$ with large variance σ^2.

that the operation can be carried out separately and independently for all locations x. In practice this is only done for a finite number of locations, by sequentially visiting the grid cells or polygons of the map. For each location $x \in D$ we have:

$$U(x) = g(A_1(x), \ldots, A_m(x)) \tag{4.2}$$

The problem now is to determine the mean $\zeta(x)$ and variance $\tau^2(x)$ of $U(x)$, or possibly the entire distribution of $U(x)$, from the operation $g(\cdot)$ and the joint distribution of the input attributes $A_i(x)$. Here it is useful to recall that, under the normal assumption, the joint distribution of the $A_i(x)$ is uniquely determined by their means $b_i(x)$, variances $\sigma_i^2(x)$ and correlations $\rho_{ij}(x, x)$. Note also that the distribution of $U(x)$ does not depend on the autocorrelations $\rho_{ii}(x, x')$ and cross-correlations $\rho_{ij}(x, x')$, for $x \neq x'$. This means that the error propagation analysis can also be carried out when the $\rho_{ii}(x, x')$ and $\rho_{ij}(x, x')$ are unknown. The situation is different when the aim is also to derive the spatial autocorrelation of $U(\cdot)$, but this will not be done here.

Because error propagation with point operations is in fact a non-spatial problem, the location index x plays a dummy role in the analysis. Therefore in this chapter and the next I shall omit the index x, to improve readability. I should also note that the solution to the problem just defined is not generally easy, and a greater part of this chapter will be devoted to it.

In figure 4.2, the problem is illustrated graphically for the one-dimensional case ($m = 1$). The figure shows that the distribution $f_U(\cdot)$ of U depends on the distribution $f_A(\cdot)$ of A as well as on the behaviour of $g(\cdot)$ near the kernel of $f_A(\cdot)$.

4.2   Four techniques of error propagation

Analytical approaches to solving the problem

The error propagation problem is relatively easy when $g(\cdot)$ is a linear function. In this case the mean and variance of U can be directly and analytically derived from the first and second moments of the A_i (Helstrom, 1991, section 4-4). The theory on functions of random variables also provides several analytical approaches to the problem for nonlinear $g(\cdot)$, but few of these can be resolved by simple calculations (Wilks, 1962, p. 59; Helstrom, 1991, section 4-3). In practice, these analytically-driven methods nearly always rely on numerical methods for a complete evaluation. Even for the relatively simple case of finding the distribution of the product of two normally distributed random variables, much computing time is required (Meeker *et al.*, 1980).

Because the goal of this research is to provide solution methods that are generally applicable, practically feasible and that do not pose strict conditions on the operation $g(\cdot)$ or on the distribution of the input attributes A_i, analytical methods are not very suitable. In this section I present four alternative methods that are more appropriate for the problem at hand. These are the first and second order Taylor methods, Rosenblueth's method and the method of Monte Carlo simulation.

4.2.1   First order Taylor method

The Taylor method is also known as the delta method (Oehlert, 1992) or as finite order analysis (Dettinger and Wilson, 1981; Kuczera, 1988). The idea of the method is to approximate $g(\cdot)$ by a truncated Taylor series centred at $\bar{b} = (b_1, \ldots, b_m)$. In the case of the first order Taylor method, $g(\cdot)$ is linearised by taking the tangent of $g(\cdot)$ in $\bar{b}$. Note that this implies that the method can only be used when $g(\cdot)$ is continuously differentiable. The linearisation greatly simplifies the error analysis, but only at the expense of introducing an approximation error.

The first order Taylor series of $g(\cdot)$ around $\bar{b}$ is given by:

$$U = g(\bar{b}) + \sum_{i=1}^{m} \{(A_i - b_i)g_i'(\bar{b})\} + \text{remainder} \tag{4.3}$$

where $g_i'(\cdot)$ is the first derivative of $g(\cdot)$ with respect to its i-th argument. The remainder of (4.3) contains the higher order Taylor terms of $g(\cdot)$, whose contribution to the result of $g(\cdot)$ is comparatively small in the neighbourhood of $\bar{b}$ (Casella and Berger, 1990, section 7.4.2). By neglecting these higher order terms the mean and variance of U are given as:

$$\zeta = E[U] \approx E\left[g(\bar{b}) + \sum_{i=1}^{m} \{(A_i - b_i)g_i'(\bar{b})\}\right] = g(\bar{b}) \tag{4.4}$$

$$\tau^2 = E[(U - E[U])^2] \approx E\left[\left(g(\bar{b}) + \sum_{i=1}^{m} \{(A_i - b_i)g_i'(\bar{b})\} - g(\bar{b})\right)^2\right]$$

$$= E\left[\left(\sum_{i=1}^{m} \{(A_i - b_i)g_i'(\bar{b})\}\right) \cdot \left(\sum_{j=1}^{m} \{(A_j - b_j)g_j'(\bar{b})\}\right)\right]$$

$$= \sum_{i=1}^{m} \sum_{j=1}^{m} \{\rho_{ij}\sigma_i\sigma_j g_i'(\bar{b})g_j'(\bar{b})\} \tag{4.5}$$

The mean of U is thus the function value of the input means b_i. In other words, it yields the same result as when the presence of errors V_i would have been completely ignored. The variance of U is the sum of various terms, which contain the correlations and standard deviations of the A_i and the first derivatives of $g(\cdot)$ at $\bar{b}$. Thus τ^2 not only depends on the variances and correlations of the input, but also on the steepness of the function $g(\cdot)$. This reflects the sensitivity of U for small changes in each of the A_i. From (4.5) it also appears that the correlations of the input errors can have a marked effect on the variance of U. Note also that (4.5) constitutes a well-known result from standard error analysis theory (Parratt, 1961; Taylor, 1982; Burrough, 1986, pp. 128–31).

For $m = 1$, (4.4) and (4.5) reduce to:

$$\zeta \approx g(b) \tag{4.6}$$

$$\tau^2 \approx \sigma^2[g'(b)]^2 \tag{4.7}$$

These results are graphically illustrated in figure 4.3. The graphs confirm that ζ equals the function value at b, and that τ^2 is larger when $g(\cdot)$ varies steeply at b.

From figure 4.3 we can also gather insight on the approximation error invoked by the Taylor method. It appears that when there is considerable variation in the derivative of $g(\cdot)$ in the region where the distribution of A has substantial support, then this can lead to a large approximation error. The

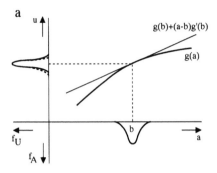

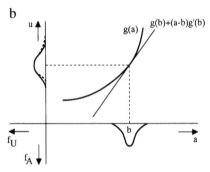

Figure 4.3   Error propagation with the first order Taylor method: (a) $g(\cdot)$ is moderate and convex upward, (b) $g(\cdot)$ is steep and concave upward. Dashed lines represent the true distribution of U.

approximation error will be negligible when $g(\cdot)$ is sufficiently smooth and/or when the input error variances are relatively small.

4.2.2   Second order Taylor method

To decrease the approximation error invoked by the first order Taylor method, one option is to extend the Taylor series of $g(\cdot)$ to include a second order term:

$$U = g(\bar{b}) + \sum_{i=1}^{m} \{(A_i - b_i)g_i'(\bar{b})\}$$

$$+ \frac{1}{2} \sum_{i=1}^{m} \sum_{j=1}^{m} \{(A_i - b_i)(A_j - b_j)g_{ij}''(\bar{b})\} + \text{remainder} \qquad (4.8)$$

where $g_{ij}''(\cdot)$ is the second derivative of $g(\cdot)$ with respect to its i-th and j-th argument. When the remainder of (4.8) is ignored, $g(\cdot)$ is approximated by a quadratic function that is locally near $\bar{b}$ a better approximation of $g(\cdot)$ than the first order Taylor series. In this case, the approximate values for the mean and variance of U are (Heuvelink *et al.*, 1989):

$$\zeta \approx g(\bar{b}) + \frac{1}{2} \sum_{i=1}^{m} \sum_{j=1}^{m} \{\rho_{ij}\sigma_i\sigma_j g''_{ij}(\bar{b})\} \tag{4.9}$$

$$
\begin{aligned}
\tau^2 \approx & \sum_{k=1}^{m} \sum_{l=1}^{m} \{\rho_{kl}\sigma_k\sigma_l g'_k(\bar{b})g'_l(\bar{b})\} \\
& + \sum_{k=1}^{m} \sum_{i=1}^{m} \sum_{j=1}^{m} \{E[(A_k - b_k)((A_i - b_i)(A_j - b_j) - \rho_{ij}\sigma_i\sigma_j)]g'_k(\bar{b})g''_{ij}(\bar{b})\} \\
& + \frac{1}{4} \sum_{i=1}^{m} \sum_{j=1}^{m} \sum_{k=1}^{m} \sum_{l=1}^{m} \{(E[(A_i - b_i)(A_j - b_j)(A_k - b_k)(A_l - b_l)] \\
& - \rho_{ij}\sigma_i\sigma_j\rho_{kl}\sigma_k\sigma_l)g''_{ij}(\bar{b})g''_{kl}(\bar{b})\}
\end{aligned}
\tag{4.10}
$$

Equation (4.9) shows that the output mean may differ from the function value of the input means. As can be seen from (4.10), to estimate the variance of U requires the third and fourth moments of the A_i. When the A_i are jointly normally distributed, these can be expressed in terms of their first and second moments (Parzen, 1962, pp. 93–5), so that (4.10) reduces to:

$$
\begin{aligned}
\tau^2 \approx & \sum_{k=1}^{m} \sum_{l=1}^{m} \{\rho_{kl}\sigma_k\sigma_l g'_k(\bar{b})g'_l(\bar{b})\} \\
& + \frac{1}{4} \sum_{i=1}^{m} \sum_{j=1}^{m} \sum_{k=1}^{m} \sum_{l=1}^{m} \{(\rho_{ik}\sigma_i\sigma_k\rho_{jl}\sigma_j\sigma_l + \rho_{il}\sigma_i\sigma_l\rho_{jk}\sigma_j\sigma_k)g''_{ij}(\bar{b})g''_{kl}(\bar{b})\}
\end{aligned}
\tag{4.11}
$$

4.2.3 Rosenblueth's method

Just as the Taylor methods, Rosenblueth's method (Rosenblueth, 1975) gives approximate solutions for the mean and variance of U, given the function $g(\cdot)$ and the first two moments of the A_i. The advantage of the method is that it is also applicable to situations in which $g(\cdot)$ is not continuously differentiable.

For the general m-dimensional case, the method estimates ζ and τ^2 from 2^m function values $g(d_k)$, $k = 1, \ldots, 2^m$, where $d_k = (d_1, \ldots, d_m)$ and each of the d_i either equals $b_i + \sigma_i$ or $b_i - \sigma_i$. The function $g(\cdot)$ is thus evaluated for all 2^m corners of the hyperquadrant that has $\bar{b}$ as its centre, and that has width $2\sigma_i$ for axis i. See Rosenblueth (1975) for a graphical illustration for $m = 2$ and $m = 3$. The mean and variance of U are then estimated from the 2^m function values as follows:

$$\zeta \approx \sum_{k=1}^{2^m} r_k g(d_k) \tag{4.12}$$

$$\tau^2 \approx \sum_{k=1}^{2^m} \left\{ r_k \left(g(d_k) - \sum_{l=1}^{2^m} r_l g(d_k)_l \right)^2 \right\} \tag{4.13}$$

where some enumeration of the 2^m points d_k has been chosen. The weights r_k are given as:

$$r_k = \frac{1}{2^m} \left(\sum_{i=1}^{m-1} \sum_{j=i+1}^{m} \delta_{ij}(k)\rho_{ij} + 1 \right) \tag{4.14}$$

where the $\delta_{ij}(k)$ depend on the coordinates of $d_k = (d_1, \ldots, d_m)$ as follows:

$$\delta_{ij}(k) = +1 \quad \text{if } d_i = b_i + \sigma_i \text{ and } d_j = b_j + \sigma_j \tag{4.15a}$$

$$\delta_{ij}(k) = +1 \quad \text{if } d_i = b_i - \sigma_i \text{ and } d_j = b_j - \sigma_j \tag{4.15b}$$

$$\delta_{ij}(k) = -1 \quad \text{if } d_i = b_i + \sigma_i \text{ and } d_j = b_j - \sigma_j \tag{4.15c}$$

$$\delta_{ij}(k) = -1 \quad \text{if } d_i = b_i - \sigma_i \text{ and } d_j = b_j + \sigma_j \tag{4.15d}$$

4.2.4 Monte Carlo method

The Monte Carlo method (Hammersley and Handscomb, 1979; Lewis and Orav, 1989) uses an entirely different approach to analyse the propagation of errors through the point operation (4.2). The idea of the method is to compute the result of (4.2) repeatedly, with input values a_i that are randomly sampled from their joint distribution. The model results form a random sample from the distribution $F_U(\cdot)$ of U, so that parameters of $F_U(\cdot)$, such as the mean ζ and the variance τ^2, can be estimated from the sample. Properties of these estimators are well known from classical sampling theory (Cochran, 1977; Lewis and Orav, 1989; Casella and Berger, 1990).

The method thus consists of the following steps:
For each location x:

1 Repeat N times:
    (a) Generate a set of realisations a_i, $i = 1, \ldots, m$.
    (b) For this set of realisations a_i, compute and store the output
        $u = g(a_1, \ldots, a_m)$.
2 Compute and store sample statistics from the N outputs u.

A random sample from the m inputs A_i can be obtained by first generating independent realisations from a univariate uniform distribution, using a suitable pseudo random number generator (Lewis and Orav, 1989; Press et al., 1992). Next these are transformed to realisations from the distributions of the A_i (Devroye, 1986). A conditioning step will have to be included when the A_i are dependent (Johnson, 1987). One attractive method for generating realisations from a multivariate Gaussian distribution uses the Cholesky decomposition of the covariance matrix (Johnson, 1987; Ripley, 1987).

Accuracy of the Monte Carlo method

Let the outcomes of N times running operation $g(\cdot)$ with the error-perturbed inputs be u_i, $i = 1, \ldots, N$. Then ζ and τ^2 can be estimated by the sample mean m_u and sample variance s_u^2:

$$m_u = \frac{1}{N} \sum_{i=1}^{N} u_i \tag{4.16}$$

$$s_u^2 = \frac{1}{N-1} \sum_{i=1}^{N} (u_i - m_u)^2 \tag{4.17}$$

The sample mean and sample variance are both unbiased and consistent estimators of ζ and τ^2. Their variances are (Lewis and Orav, 1989, p. 123):

$$\text{var}(M_u) = \frac{\tau^2}{N} \tag{4.18}$$

$$\text{var}(S_u^2) = \frac{1}{N}\left[\mu_4 - \tau^4\left(\frac{N-3}{N-1}\right)\right] \approx \frac{\mu_4 - \tau^4}{N} \tag{4.19}$$

where μ_4 is the fourth central moment of U. The sample skewness and kurtosis can also be computed, yielding unbiased estimates of the skewness and kurtosis of U. The variances of these estimators can also be derived, although the resulting expressions are somewhat complicated (Lewis and Orav, 1989, p. 123).

When U is normally distributed, (4.19) reduces to:

$$\text{var}(S_u^2) = \frac{2\tau^4}{N-1} \tag{4.20}$$

An interesting consequence from (4.19) is that the coefficient of variation of S_u^2 is independent of τ^2:

$$\text{cv}(S_u^2) = \sqrt{\left(\frac{\gamma_2}{N} + \frac{2}{N-1}\right)} \tag{4.21}$$

where $\gamma_2 = \mu_4/\tau^4 - 3$ is the coefficient of kurtosis, which is zero for the normal distribution. If the coefficient of variation of the resulting sample variance is taken as a criterion for establishing its accuracy, then (4.21) thus enables one to tell in advance how large the number of Monte Carlo runs should be. For example, when γ_2 is zero and when the relative error in S_u^2 should be 0.05 or less, then N should be 801 or larger.

From (4.18) and (4.19) it follows that the standard deviations of M_u and S_u^2 are approximately inversely related to the square root of the number of Monte Carlo runs N. This is a general result, that holds for other parameters of the distribution as well. It means that to double the accuracy, four times as many Monte Carlo runs are needed. The accuracy thus slowly progresses as N increases. Nonetheless, the distribution of U can be approximated to any level of precision, by repeating the Monte Carlo simulation a sufficiently large number of times.

Estimation of quantiles and percentiles

An important advantage of the Monte Carlo method is that it allows not only the estimation of the mean and variance of U, but moreover its entire distribution. For instance, the median of U can be estimated directly from the sample median. More generally, estimates of quantiles and percentiles of U can be easily obtained after the N simulated values $u_i(i = 1, \ldots, N)$ have been arranged in increasing order. Let the *quantile* q_α be defined as the quantity for which $F_U(q_\alpha) = \alpha$, and let the *percentile* p_v be the quantity that satisfies $p_v = F_U(v)$ for a given value v. If the sample is ordered such that $u_i \geq u_j$ whenever $i > j$, then q_α is estimated as (Lewis and Orav, 1989, p. 148):

$$\hat{q}_\alpha = u_{\lceil \alpha N \rceil} \tag{4.22}$$

where $\lceil \alpha N \rceil$ is the smallest integer larger or equal to αN. An unbiased estimate of p_v is given by (Lewis and Orav, 1989, p. 131):

$$\hat{p}_v = \frac{1}{N} \max\{i \mid u_i \leq v\} \tag{4.23}$$

Quantiles and percentiles are particularly useful to provide information about the behaviour of U near the tails of its distribution, which is important for many applications, such as in soil pollution research (Leenaers *et al.*, 1991). For instance, let U refer to a risk factor computed from the concentrations of heavy metals in the soil (Heuvelink *et al.*, 1989). If c is the critical risk value above which sanitation measurements have to be taken, then $1 - p_c$ is the probability that the actual risk factor U exceeds the critical value. A sensible decision may then be to clean up the soil at all locations where $1 - p_c$ is greater than 0.05.

The variance of the sample quantiles and percentiles is also inversely related to N (Lewis and Orav, 1989, pp. 130 and 152). The resulting expressions are often complicated, but an exception is the variance of $\hat{p}_v$:

$$\mathrm{var}(\hat{p}_v) = \frac{p_v - p_v^2}{N} \tag{4.24}$$

4.2.5 Evaluation and comparison of the four error propagation techniques

First order Taylor method
Implementation of the first order Taylor method may seem arduous at first because (4.5) involves m^2 summations and requires the first derivatives of $g(\cdot)$. But in fact, when the method is to be compared with the other three error

propagation methods, its most important advantage is its modest computational load. This is partly because the other methods can be even more time consuming, but also because the evaluation of (4.5) is not as laborious as it may seem at first sight. Firstly, the number of summations can often be substantially reduced, due to redundancy and zero correlations. Secondly, the derivatives of $g(\cdot)$ can be obtained by means of symbolic differentiation (Wesseling and Heuvelink, 1991). This is very efficient particularly for spatial point operations, because the analytical differentiation of $g(\cdot)$ need only be carried out once, whereas numerical differentiation needs to be repeated for all locations in D (e.g. all grid cells or polygons).

The first order Taylor method is also attractive because it yields an analytical expression for the variance of the output error, although it should be remembered that the solution is approximate only. The means, variances and correlations of the input attributes explicitly appear in (4.5), and this allows one quickly to examine how τ^2 changes under changes in the input error parameters. Note also that the method does not require the joint distribution of the A_i to be known, but only their first and second moments. This is sometimes seen as an advantage, but in fact it merely demonstrates that the method ignores the influence of higher order moments.

The first order Taylor method clearly also has disadvantages. The main disadvantage is that it is an approximate method only. When the operation $g(\cdot)$ is strongly nonlinear in $\bar{b}$, then the approximation error may become unacceptably large (Kuczera, 1988). On the other hand, the approximation error is zero when $g(\cdot)$ is linear, so that (4.4) and (4.5) contain the exact result for linear $g(\cdot)$. Another disadvantage of the first order Taylor method is that it requires that $g(\cdot)$ is continuously differentiable, and so it does not apply to operations such as Maximise and Minimise.

Although ease of computation was mentioned as the main advantage of the first order Taylor method, this may no longer be true when the operation $g(\cdot)$ is a complicated computational model that contains many inputs. Although such models can still be dealt with by the Taylor method, it is often difficult to keep track of the approximation errors involved (Jones, 1989).

Second order Taylor method

Because the second order Taylor method is a natural extension of the first order Taylor method, many of the advantages and disadvantages given above also apply to the second order method.

The difference between the first and second order methods is that the latter generally yields smaller approximation errors, at the expense of a considerable increase of numerical complexity. Although the second order approximation of $g(\cdot)$ is always better in the neighbourhood of $\bar{b}$, the approximation may actually get worse further away from $\bar{b}$. Consequently, it is unjust to conclude that the results obtained with the second order method will always be more accurate, as suggested by Dettinger and Wilson (1981). In view of the fact that the evaluation of (4.10) or (4.11) is much more laborious than the evaluation of (4.5), there will be few practical situations in which the second order method is to be preferred over the first order method. One important exception is when $g(\cdot)$ is a quadratic function, in which case the second order method is error-free and the first order method is not. Sections 5.1 and 5.3 give examples.

Rosenblueth's method

Rosenblueth's method should be interpreted as an alternative to the first order Taylor method. The method is also exact when $g(\cdot)$ is linear, but approximation errors will generally result when $g(\cdot)$ is nonlinear.

To understand the difference between the first order Taylor and Rosenblueth's method, consider the one-dimensional situation. For $m = 1$, (4.12) and (4.13) reduce to:

$$\zeta \approx \tfrac{1}{2}(g(b + \sigma) + g(b - \sigma)) \tag{4.25}$$

$$\tau^2 \approx \tfrac{1}{4}(g(b + \sigma) - g(b - \sigma))^2 \tag{4.26}$$

When these are compared with (4.6) and (4.7), we see that whereas the first order Taylor method uses the derivative of $g(\cdot)$ at b to approximate $g(\cdot)$, Rosenblueth's method uses the function values of $g(\cdot)$ a distance $\pm\sigma$ away from b, yielding a 'smoothed' derivative of $g(\cdot)$.

Rosenblueth's method is thus a valuable alternative to the first order Taylor method when $g(\cdot)$ is not differentiable in $\bar{b}$. From a numerical perspective the method is more demanding than the first order Taylor method, especially when m is large.

Monte Carlo method

The most important advantage of the Monte Carlo method is that it can yield the entire distribution of U at an arbitrary level of accuracy. Important advantages are also that the method is easily implemented and generally applicable. Implementation is easy because the method is not affected by the exact formulation of the operation $g(\cdot)$. The method merely treats $g(\cdot)$ as a black box, whose response to the perturbed inputs is studied from the resulting outputs. Therefore, whether the operation $g(\cdot)$ is a simple empirical model relating soil hydraulic parameters to particle size distributions (Cosby et al., 1984) or a complicated dynamic conceptual model describing forest hydrological processes (Bouten, 1993), the steps of the Monte Carlo error analysis effectively remain the same.

The Monte Carlo method has also several disadvantages. The first is that its results do not come in a nice analytical form. For instance, the resulting equations of the Taylor methods may be analysed to see how a reduction of input error will have an effect on the output. With the Monte Carlo method the only solution is to run the entire simulation again.

The main disadvantage of the Monte Carlo method is its numerical load, because the operation $g(\cdot)$ must be executed N times. In most practical situations N will take a value in between 50 and 2000, but it may occasionally be as large as 100 000. When $g(\cdot)$ is a simple operation this may not be too disturbing, but what if $g(\cdot)$ represents a complex computational model that itself takes much computing time? The total computing time required may then become formidable.

The problem of the heavy computational load also draws attention to the question of how large N should be. In groundwater modelling practice, it has been stated that $N = 100$ is sufficient to obtain a reasonable estimate of the mean ζ, $N = 1000$ is the minimum required for the variance τ^2, and to estimate the 1 per cent quantile requires tens of thousand of runs (Peck et al., 1988, pp. 129–30). In practice, the number of runs rarely exceeds 500 (Smith and Freeze, 1979; Smith and Hebbert, 1979; Jones, 1989; De Roo et al., 1992; Bierkens and Burrough, 1993b). In GIS research, studies are carried out that use the Monte Carlo method with only 20 or even 10 runs (Fisher, 1992; Goodchild et al., 1992), although Goodchild et al. (1992) acknowledge that 10 runs are not sufficient to obtain accurate estimates of the mean. Openshaw (1989) claims that in practice, remarkably few runs are needed to evaluate the propagation of error in GIS, but he does not support this statement with arguments. Using only 100 runs to estimate the 1 per cent quantile (Openshaw et al., 1991) seems much more to be guided by limitations of computing capacity than by conscientious accuracy considerations.

4.3 Error propagation with neighbourhood operations

There is a simple way to extend the results for the point operation to neighbourhood operations. Neighbourhood operations can be reformulated as point operations, by defining new attributes that are shifted versions of the original attribute. The idea can best be demonstrated by an example. Suppose a remote sensing image is smoothed using a 3 by 3 low pass filter (Curran, 1985):

1/9	1/9	1/9
1/9	1/9	1/9
1/9	1/9	1/9

The output is thus obtained by averaging the values of the input attribute at the 9 grid cells that surround and include the cell of interest. This problem can also be interpreted as one in which 9 input maps are averaged, where each of the 9 maps is derived from the original map using a suitable translation. For example, the contribution of the upper left corner cell is given by the map that is obtained from moving the original map one cell to the right and one cell downwards.

A neighbourhood operation can thus be redefined as a point operation, at the expense of increasing the number of inputs m. The propagation of errors can then be analysed using the methods described in the previous section. The only difference is that the correlations between the input attributes must be obtained from the autocorrelation of the original attribute.

In the case of the 3 by 3 low pass filter, which is a linear operation, the error propagation can best be analysed using the first order Taylor method. Careful analysis of (4.5) then shows that the smoothing has the most effect when spatial autocorrelation is absent, in which case the error variance is reduced by a factor of 9 (assuming $\sigma^2(\cdot)$ spatially invariant). This result agrees with the discussion in section 2.5, where it was found that error is generally smaller for larger blocks, especially when much of the within-block variability is averaged out.

When the neighbourhood increases, in principle the error analysis can still be carried out in the way described above, but the problem is that the number of input layers m will increase dramatically. For instance, a 5 by 5 window results in 25 input layers. This causes problems of numerical complexity, so that other approaches become more suitable. I will discuss these in chapter 6, together with the general problem of error propagation with global GIS operations.

4.4   Sources of error contributions

When the error analysis reveals that the output of $g(\cdot)$ is insufficiently accurate then measures will have to be taken to improve accuracy. When there is a single input to $g(\cdot)$ then there is no doubt where the improvement must be sought, but what if there are multiple inputs to the operation? Also, how much should the error of a particular input be reduced in order to reduce the output error by a given factor? These are important questions that will be considered in this section.

At this point I should note that although these questions are addressed before I consider the propagation of errors with global operations, this does not mean that the results of this section are only valid for local operations. Many of the results given below can be generalised to include global operations as well.

The partitioning property

To obtain answers to the questions above, consider (4.5) again, which gives the variance of the output U using the first order Taylor method:

$$\tau^2 \approx \sum_{i=1}^{m} \sum_{j=1}^{m} \{\rho_{ij}\sigma_i\sigma_j g_i'(\bar{b})g_j'(\bar{b})\} \tag{4.27}$$

When the inputs are uncorrelated, this reduces to:

$$\tau^2 \approx \sum_{i=1}^{m} \{\sigma_i^2 (g_i'(\bar{b}))^2\} \tag{4.28}$$

Equation (4.28) shows that the variance of U is the summation of parts, each to be attributed to one of the inputs A_i. This *partitioning property* allows one to analyse how much each input contributes to the output variance. Thus from (4.28) it can directly be seen how much τ^2 will reduce from a reduction of σ_i^2. Clearly the output will mainly improve from a reduction in the variance of the input that has the largest contribution to τ^2. Note that this need not necessarily be the input with the largest variance, because the operation $g(\cdot)$ need not be very sensitive for the input with the largest variance.

When inputs are correlated there are also mixed terms in the output variance (4.27). These cannot be attributed to a single input, but they are the result of the interactions between two inputs. These mixed terms should be included in an account of the joint contribution of input errors. Note that mixed terms may well be negative, such as is the case when two negatively correlated attributes are added or when two positively correlated attributes are subtracted.

Note that (4.27) and (4.28) are only approximate when $g(\cdot)$ is nonlinear. In such a situation more accurate results can be obtained with the Monte Carlo method. This can be done by simply setting all input errors to zero, except those of interest. The resulting output variance should then be compared with the output variance that was obtained when all error sources were included. A more efficient method, though still computationally demanding, is to use a clever random sampling design in the Monte Carlo method (Jansen *et al.*, 1994).

Incorporation of model error

As stated in the introductory chapter, a GIS operation is often in effect a computational model. Consequently, not only will *input error* propagate to the output of a GIS operation, but *model error* will as well. This means that the output of a computational model may disagree with reality even when the input to the model is completely error-free, simply because the model is only an approximation of reality. In many situations model error will be a major source of error and should therefore be included in the error analysis. Ignoring it would severely underestimate the true uncertainty in the model output.

Model error is usually included by assigning errors to model coefficients and/or by adding a residual error term to the model equation(s). A typical example is an empirical regression model, where both sorts of error are present (Webster and Oliver, 1990, p. 99). Regression coefficients are estimated from a finite sample and so they are subject to sampling error. A residual error occurs because rarely will all the variability in the dependent variable be accounted for by the regression.

The user of an empirical model should be aware of the uncertainties to be associated with the model. Unfortunately, it happens too often that empirical models are published with absolutely no information about their accuracy. And even when accuracy estimates are given there is still cause for concern because an empirical model has no universal accuracy. For example, an empirical erosion model developed and calibrated in one area may perform much worse when it is applied to another area.

Clearly conceptual (physically based) models also suffer from errors. These errors are caused by the various assumptions, discretisations and simplifications that are purposely made to make the model manageable (Gilchrist, 1984; Van Geer *et al.*, 1991; Beven and Binley, 1992; Grayson *et al.*, 1992a,b; Van Der Perk, 1997). In many cases the error in a conceptual model will be represented by an additive noise term (Kuczera, 1988), also referred to as the system noise (Van Geer *et al.*, 1991), which can be estimated by means of validation (Willmott, 1981; Luis and McLaughlin, 1992; Heuvelink, 1998; Van Der Perk, 1997). Note also that model error may also include a systematic component (Moore and Rowland, 1990; Van Geer *et al.*, 1991).

Once model error has been quantified, it can easily be incorporated in the error analysis by augmenting the number of stochastic inputs to the operation $g(\cdot)$.

The balance of errors

The original problem stated at the beginning of this chapter was to determine how large the errors in the output of a GIS operation are, given the errors in the input and the kind of operation used. This problem can be approached using one of the techniques that I discussed in the previous section. However, it is important to note that an error analysis offers much more than only the computation of output error.

An error analysis can also determine how much each individual input contributes to the output error. This information may be very useful, because it allows users to explore how much the quality of the output improves, given a reduction of error in a particular input. Note that when the spatial variability is known such a cost–benefit analysis can be performed beforehand, because the reduction in the kriging prediction error can be computed before the additional observations are made (McBratney *et al.*, 1981; Webster and Oliver, 1990; Burrough, 1991). Thus the improvement foreseen due to intensified sampling can be weighed against the extra sampling costs.

If a reduction of output error is required, it need not necessarily be sensible to improve the input with the highest error contribution. This is because some input attribute errors may be more expensive to measure than others. However, in many cases it will be most rewarding to strive for a *balance of errors*. When the error in an attribute has a marginal effect on the output, then there is little to be gained from mapping it more accurately. In that case, extra sampling efforts can much better be directed to an input attribute that has a larger contribution to the output error. For instance, if a pesticide leaching model is sensitive to soil organic carbon and less so to soil bulk density, then it is more important to map the former more accurately (Loague *et al.*, 1989).

The example of the pesticide leaching model draws attention to the fact that a balance of errors must also include model error. It is clearly unwise to spend much effort on collecting data if what is gained is immediately thrown away by using a poor computational model. On the other hand, a simple computational model may be as good as a complex model if the latter needs lots of data that cannot be accurately obtained. This is why many researchers in catchment hydrology raise the question of whether there is much benefit to be gained from developing ever more complex models when the necessary inputs cannot be evaluated in the required spatial and temporal resolution (Beven, 1989; Grayson *et al.*, 1992b; De Roo, 1993).

It is important to add that with the advent of GIS, and the many computational models that often come freely with it, there is an increased risk of disturbing the balance between input and model error. When there is no protection against improper use then ignorant users will be tempted to apply models to scales or use them for purposes for which they were not developed (Heuvelink, 1998). It is a clear form of ill-use when the GIS is used to link and combine models with different levels of precision and resolution. These problems can only be tackled when users become aware of them and when error propagation is explicitly incorporated in the analysis (Heuvelink *et al.*, 1989; Burrough, 1992b).

CHAPTER FIVE

Error propagation with local GIS operations: applications

This chapter demonstrates how the theory presented in the previous chapters can be applied to relevant practical problems. Four case studies are considered, in increasing order of complexity. The aim of this chapter is not so much to thoroughly describe the theory behind the four case studies or to analyse all aspects of the problems at hand, but to show how error propagation theory can be applied in practice. Thus the case studies are used to show how to apply an error analysis and how meaningful conclusions can be drawn from it.

5.1 Predicted lead consumption in the Geul river valley

The floodplain of the Geul valley, located in the south of the Netherlands, is strongly polluted by heavy metals deposited with the stream sediments. Historic metal mining has caused the widespread dispersal of lead, zinc and cadmium in the alluvial soil. Figure 5.1 shows a map of the study area. The pollutants may constrain the land use in these areas, so detailed maps are required that delineate zones with high concentrations. Leenaers (1991) demonstrated that accurate pollution maps can be produced by either kriging from point data on soil metal concentrations or cokriging from additional elevation data. Using point kriging on measurements of soil samples of 100 grams at 101 sites in the area, maps were derived of the predicted lead concentration in the topsoil (0–10 cm depth) and of the standard deviations of the prediction errors. The variogram that was obtained from the observations and used for kriging was a spherical model with a nugget of 5550 $(mg/kg)^2$, a sill of 27450 $(mg/kg)^2$ and a range of 0.397 km. The maps are given in figure 5.2. In the analysis hereafter it is assumed that the kriging prediction error of lead is normally distributed.

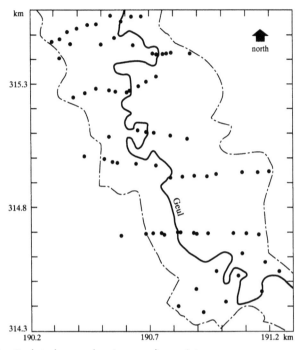

Figure 5.1   The Geul study area showing sampling points.

For the general assessment of health risks to children playing in the Geul valley, it could be sensible to make maps of potential daily lead ingestion. This can be done if it is known how much soil a child is likely to ingest per day. Van Wijnen *et al.* (1990) provide experimental data on the daily ingestion of soil by young children playing on a camping ground. After making correction

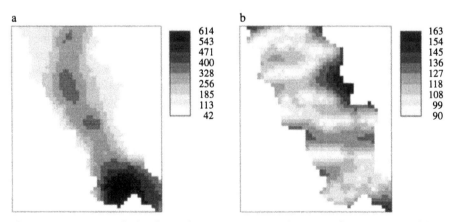

Figure 5.2   Kriging results for the Geul area (25 × 25 m grid): (a) point kriging mean and (b) standard deviation of lead concentration of topsoil (mg/kg).

for 'background' ingestion, these data fit a lognormal distribution with mean 0.120, median 0.052 and standard deviation of 0.250 g/day. Maps of the potential daily ingestion of lead (I) can now be obtained by multiplying the lead concentration of the soil (PB) by the amount of soil consumed (S). Uncertainty in both the lead concentration of the soil and the daily soil ingestion will propagate to the daily lead ingestion. This implies that even when a site is on average safe, there can be incidences in which the children's health is at risk. The propagation of the errors will now be analysed using all four error propagation techniques.

Error analysis with the second order Taylor method

Since the operation considered is the multiplication of two attributes, the second order Taylor method is appropriate here. The errors in soil ingestion and lead concentration of soil are uncorrelated, and so (4.9) and (4.10) yield in this case:

$$\text{mean}(I) = \text{mean}(PB) \cdot \text{mean}(S) \tag{5.1}$$

$$\text{var}(I) = \text{var}(PB) \cdot (\text{mean}(S))^2 + \text{var}(S) \cdot (\text{mean}(PB))^2 + \text{var}(PB) \cdot \text{var}(S) \tag{5.2}$$

There is equality in (5.1) and (5.2) because the remainder of (4.8) is zero for a quadratic function. The mean and standard deviation of the daily lead ingestion are given in figure 5.3. Note the large values for the standard deviation in figure 5.3b.

Because the second order Taylor method is exact, its results can now be used to investigate the performance of the first order Taylor method and

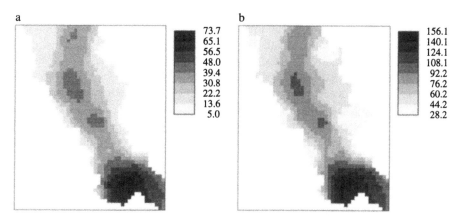

Figure 5.3 Error propagation results for potential daily lead ingestion (μg/day) in the Geul valley: (a) mean and (b) standard deviation as obtained with the second order Taylor method.

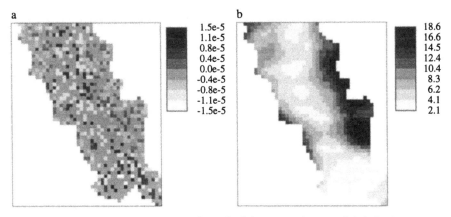

Figure 5.4 Differences in the computed standard deviation of potential daily lead ingestion (μg/day): (a) second order Taylor minus Rosenblueth, (b) second order Taylor minus first order Taylor.

Rosenblueth's method. Comparison of (4.4), (4.9) and (4.12) shows that in this case both methods exactly reproduce the mean of the potential daily soil ingestion. Figure 5.4a shows that Rosenblueth's method also reproduces the standard deviation of the potential daily lead ingestion, which may be verified from comparison of (4.10) and (4.13). The first order Taylor method causes moderate approximation errors (figure 5.4b).

Error analysis with the Monte Carlo method

Although Rosenblueth's method and the second order Taylor method are exact, their disadvantage is that they only yield the mean and standard deviation of the potential daily lead ingestion, whereas the main interest lies in whether there is a substantial risk that critical values of lead ingestion are exceeded. In this case the critical value is given by the Acceptable Daily Intake (ADI), which for lead is given by 50 μg/day (Leenaers et al., 1991). The Monte Carlo method can now be employed to obtain the percentiles of I for this critical value. In figure 5.5 maps are given of the probability that the actual lead ingestion is larger than the ADI of 50 μg/day.

The number of Monte Carlo runs was taken as $N = 1000$ and $N = 10\,000$. Figure 5.5 shows that $N = 1000$ is not yet sufficient to reach stable results, but $N = 10\,000$ seems sufficient. The accuracy of these maps can be analysed with (4.24). For example, when the percentile p_{ADI} equals 0.950 for some location, then the standard deviation of its estimator $\hat{p}_{ADI}$ is 0.0069 for $N = 1000$ and 0.0022 for $N = 10\,000$.

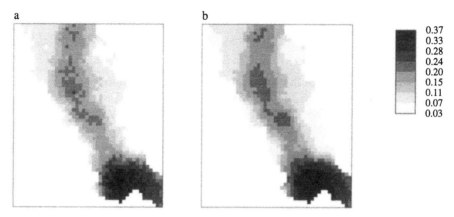

Figure 5.5 Probability of exceeding the Acceptable Daily Intake for lead (50 μg/day): (a) Monte Carlo with 1000 runs, (b) Monte Carlo with 10 000 runs.

Support size problems

Analysis of (5.2) shows that the large uncertainty in the potential daily lead ingestion is mainly caused by the uncertainty in the daily soil consumption. This would suggest that there is not much benefit to be expected from a more accurate mapping of the lead concentration of the soil. However, if the interest goes to the potential lead ingestion over a longer period of time, this conclusion may no longer be valid. This is because an increase of support in time, for instance to a week instead of one day, will decrease the variability in soil consumption, whereas the uncertainty in lead concentration remains the same. It is difficult to predict how much of the variability in soil consumption averages out over a longer period of time, because the variability in children's behaviour remains. The change of support problem can better be avoided by measuring the soil ingestion for the time support that one is interested in.

There are even larger problems with respect to which soil support to use in the analysis. In this example point kriging was used, and so the support used refers to the size of the soil samples, which are about 100 grams. The average daily soil consumption is much less, which would suggest that the variability in the lead concentration of consumed soil can be much larger than given in figure 5.2b. However, one should also bear in mind that the daily soil ingestion is not a single soil sample taken from one location only. It is the sum of many small portions taken at various locations on the camping ground, so variation averages out. It is hard to tell how these counteracting effects combine to form the true variability in the lead concentration of consumed soil.

These problems with change of support clearly show that the results of the error analysis must be treated with care. Any user of the figures presented here should be aware that the results are only valid for the support that was used to collect the data.

5.2 Slope and aspect of the Balazuc digital elevation model

Figure 5.6 shows a digital elevation model (DEM) of the Balazuc area in France, which is 4.6 × 3.2 km in size. The map was obtained by interpolating a digitised contour map to a grid of 25 × 25 m (De Jong and Riezebos, 1988). In this particular case it proved to be impossible to recapture how the various sources of error, i.e. the errors in the contour map itself, the digitising error and the interpolation error, combine to yield an overall error map of the DEM. Therefore an error map could not be calculated but was gathered from the experience of the people that constructed the DEM. Here it will be assumed that the error in the point elevation is normally distributed with zero mean and standard deviation equal to 5 m, independent of the position in the area. Clearly these informed guesses are no substitute for real values, and so it is important that DEM products are accompanied by a measure of their accuracy, such as the root mean square error, RMSE (Fisher, 1992).

Maps of the slope and aspect of the Balazuc area will now be calculated using the second order and third order finite difference methods (Skidmore, 1989). The third order method uses data from a larger window and so it yields more smoothed results. Because of this it is generally found to be more accurate than the second order method (Skidmore, 1989; Carter, 1992), although one should be aware that the smoothing causes finer details to be lost (Carter, 1992). The purpose of the present analysis is to quantify the difference in accu-

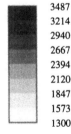

	3487
	3214
	2940
	2667
	2394
	2120
	1847
	1573
	1300

Figure 5.6   Digital elevation model of the Balazuc area (25 × 25 m grid).

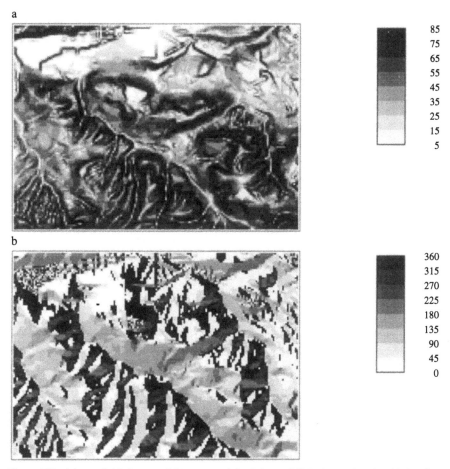

Figure 5.7   Mean of: (a) slope and (b) aspect of the Balazuc DEM when using the third order finite difference method and correlation parameter $q = 0.5$.

racy between the two methods. This will be done for three different levels of spatial autocorrelation.

Knowledge about the spatial autocorrelation is needed here because the slope and the aspect are obtained from neighbourhood operations and so they involve spatial interactions within a small window. Although it is in this case not possible to give an exact value or even a reliable estimate, it is reasonable to assume that the errors in the Balazuc DEM are positively correlated at short distances. The error analysis will be carried out using an exponential autocorrelation function:

$$\rho(x,x') = q \cdot \exp\left(-\frac{|x - x'|}{r}\right) \quad \text{for } x \neq x' \tag{5.3}$$

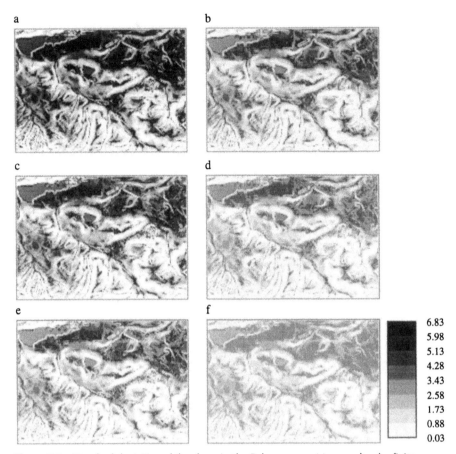

Figure 5.8 Standard deviation of the slope in the Balazuc area: (a) second order finite difference method with correlation parameter $q = 0$, (b) third order method with $q = 0$, (c) second order method with $q = 0.5$, (d) third order method with $q = 0.5$, (e) second order method with $q = 0.8$, (f) third order method with $q = 0.8$.

The range parameter r is taken as 200 metres, and the three different values for q that are used are $q = 0$, $q = 0.5$ and $q = 0.8$. These figures are chosen arbitrarily and are not supported by experimental evidence. Note that the range parameter is large with respect to the window size.

Results

The analysis was performed using the Monte Carlo method with 5000 runs ($N = 5000$). The Taylor method is not suitable here because calculating slope and aspect involves operations that are not continuously differentiable everywhere. Rosenblueth's method is not very attractive either, because it involves

a   b

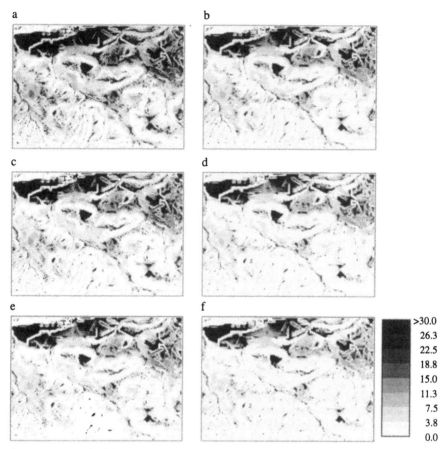

c   d

e   f

>30.0
26.3
22.5
18.8
15.0
11.3
7.5
3.8
0.0

Figure 5.9 Standard deviation of the aspect in the Balazuc area: (a) second order finite difference method with correlation parameter $q = 0$, (b) third order method with $q = 0$, (c) second order method with $q = 0.5$, (d) third order method with $q = 0.5$, (e) second order method with $q = 0.8$, (f) third order method with $q = 0.8$.

approximation errors and it is slow when the number of inputs (here $m = 4$ for the second order and $m = 8$ for the third order method) is large. Figure 5.7 gives the mean of slope and aspect using the third order finite difference method and $q = 0.5$. These maps are not very much different when the second order finite difference method or other values of q are used.

More important here are the resulting standard deviations in slope and aspect. Figure 5.8 gives the results for the slope. It appears that standard deviations can be quite large, and are indeed on average smaller for the third order finite difference method. Moreover, the accuracy in the slope increases with an increase of spatial autocorrelation. This is not surprising because smaller

errors in slope occur when the errors in elevation are more positively correlated. Because in this case the exact correlation in the DEM error is unknown,
using a zero correlation can be considered a worst case approach (Heuvelink
et al., 1990; Hunter and Goodchild, 1997).

Similar results are obtained for the aspect, see figure 5.9. The mean and
standard deviation of the aspect were computed using the formulas that are
appropriate for directional data (Mardia, 1972, section 2.4). The error in the
aspect is extremely large in flat areas. This is not surprising because in flat
areas small differences in elevation generate large deviations in aspect. Particularly, the slope is sensitive to the degree of spatial autocorrelation. It can
therefore be concluded that the error information about a DEM should not
only contain a standard deviation or RMSE, but also an estimate of the
spatial autocorrelation of the elevation errors (Goodchild *et al.*, 1992).

5.3 Predicting soil moisture content with linear regression for the Allier floodplain soils

As part of a research study in quantitative land evaluation, the WOFOST
crop simulation model (Van Diepen *et al.*, 1989) was used to calculate potential crop yields for floodplain soils of the Allier river in the Limagne rift valley,
central France. The WOFOST model was run at a number of sample sites
from which maps of the variation of potential yield were obtained by kriging
(Weterings, 1988; Stein *et al.*, 1989). The moisture content at wilting point
(Θ_{wp}) is an important input attribute for the WOFOST model. Because Θ_{wp}
varies considerably over the area in a way that is not linked directly with soil
type, it was necessary to map its variation separately to see how moisture
limitations affect the calculated crop yield.

Unfortunately, because Θ_{wp} must be measured on samples in the laboratory, it is expensive and time-consuming to determine it for a sufficiently
large number of data points for kriging. An alternative and cheaper strategy is
to calculate Θ_{wp} from other attributes that are cheaper to measure. Because
the moisture content at wilting point is often strongly correlated with the
moisture content at field capacity (Θ_{fc}) and the bulk density (BD) of the soil –
both of which can be measured more easily – it was decided to investigate how
errors in measuring and mapping these would work through to a map of
calculated Θ_{wp}. In this case study I report the results in terms of porosity (Φ),
which is linearly derived from bulk density as $\Phi = (1 - BD/2.65)$.

The following procedure, illustrated in figure 5.10, was used to obtain a
map of the mean and standard deviation of Θ_{wp}.

1 The properties Θ_{wp}, Θ_{fc} and Φ were determined in the laboratory for
   100 cm^3 cylindrical samples taken from the topsoil (0–20 cm) at 12 selected
   sites shown as the circled points in figure 5.11. The results are given in table
   5.1.

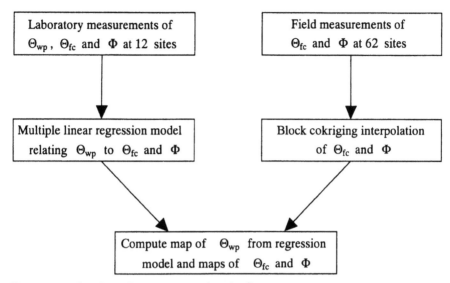

Figure 5.10   Flowchart of mapping procedure for Θ_{wp}.

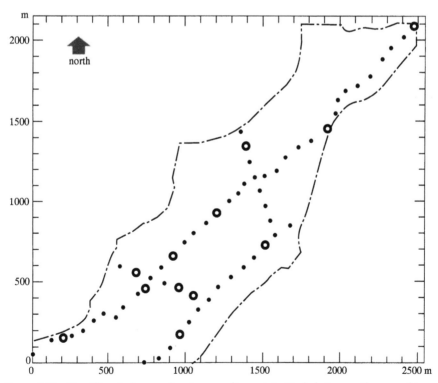

Figure 5.11   The Allier study area showing sampling points. Circled sites are those used to estimate the regression model.

2 These results were used to set up a pedotransfer function, relating Θ_{wp} to Θ_{fc} and Φ, which took the form of a multiple linear regression:

$$\Theta_{wp} = \beta_0 + \beta_1\Theta_{fc} + \beta_2\Phi + \varepsilon \tag{5.4}$$

The coefficients β_0, β_1 and β_2 were estimated using standard ordinary least squares regression. The estimated values for the regression coefficients and their respective standard deviations were $\hat{\beta}_0 = -0.263 \pm 0.031$, $\hat{\beta}_1 = 0.408 \pm 0.096$, $\hat{\beta}_2 = 0.491 \pm 0.078$. The standard deviation of the residual ε was estimated as 0.0114. The correlation coefficients of the regression coefficients were $\rho_{01} = -0.221$, $\rho_{02} = -0.587$, $\rho_{12} = -0.655$. The goodness-of-fit of the regression (R^2 is 94.8 per cent) indicates that the model is satisfactory (see figure 5.12). Note that the presence of spatial correlation between the observations at the 12 locations was ignored in the regression analysis.

3 Sixty-two measurements of Θ_{fc} and Φ were made in the field at the sites indicated in figure 5.11. From these data experimental variograms were computed. These were then fitted using the linear model of coregionalisation (Journel and Huijbregts, 1978). Table 5.2 gives the parameters of the fitted model. Thus each variogram is a weighted sum of a nugget variogram, a spherical variogram with a range of 500 m and a spherical variogram with a range of 800 m. Note the negative nugget of the cross-variogram. For the purposes of this study the input data for the regression were mapped to a regular 50 × 50 m grid using block cokriging with a block size of 50 × 50 m. The block cokriging yielded raster maps of means and standard deviations for both Θ_{fc} and Φ, as well as a map of the correlation of the block cokriging prediction errors (see section 2.4). The geostatistical software package gstat (Pebesma and Wesseling, 1998) has the functionality to

Table 5.1 Laboratory results of Θ_{wp}, Θ_{fc} and Φ (cm^3/cm^3) at 12 selected sites (0–20 cm).

Point	Θ_{wp}	Θ_{fc}	Φ
1	0.072	0.272	0.419
2	0.129	0.369	0.491
3	0.189	0.392	0.566
4	0.103	0.334	0.464
5	0.086	0.304	0.453
6	0.114	0.328	0.532
7	0.205	0.363	0.634
8	0.199	0.451	0.566
9	0.108	0.299	0.509
10	0.103	0.337	0.491
11	0.112	0.318	0.509
12	0.103	0.337	0.479

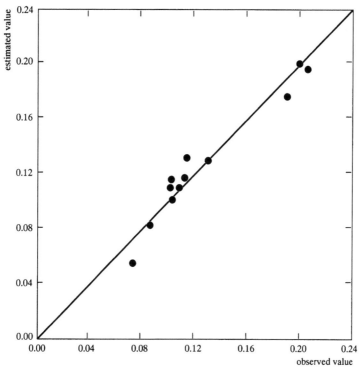

Figure 5.12   Observed values versus estimated values of Θ_{wp} for the 12 selected sites.

compute the map of correlations. Figure 5.13 displays these maps. Note that there are clear spatial variations in the correlation between the block cokriging errors.

4 The maps of Θ_{fc} and Φ were substituted in the regression equation (5.4), yielding a map of the attribute Θ_{wp}. To determine whether the accuracy of this map would be acceptable an error propagation analysis was carried out, using the second order Taylor method. Because the inputs to the model are averages for blocks that are much larger than the sample support, it was assumed that the residuals of the regression cancel out. The residual error ε

Table 5.2   Parameters of the variograms and cross-variogram for Θ_{fc} and Φ as fitted using the linear model of coregionalisation.

	Nugget	Spherical range 500 m	Spherical range 800 m
Variogram of Θ_{fc}	0.0023	0.0015	0.0003
Variogram of Φ	0.0020	0.0003	0.0017
Cross-variogram of Θ_{fc} and Φ	−0.0005	0.0005	0.0006

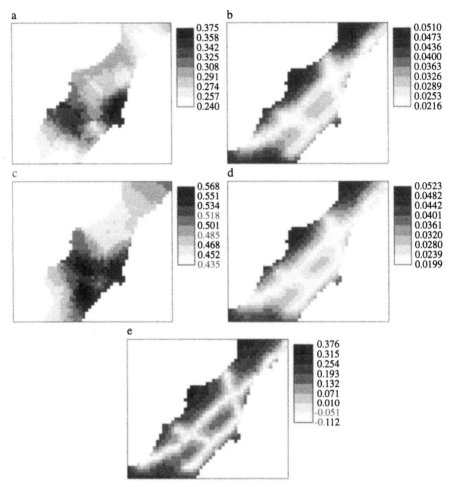

Figure 5.13 Kriging results for the Allier study area (50 × 50 m grid): (a) block mean and (b) standard deviation of soil moisture content at field capacity Θ_{fc} (cm³/cm³), (c) block mean and (d) standard deviation of soil porosity Φ (cm³/cm³), (e) correlation of cokriging prediction errors of Θ_{fc} and Φ.

was therefore not included in the error analysis. Because the model coefficients and the field measurements were determined independently, the correlation between the $\hat{\beta}_i$ and the cokriging prediction errors was taken to be zero. The maps of the mean and standard deviation of Θ_{wp} were then computed with (4.9) and (4.11). Note that to use (4.11) it was assumed that the errors are jointly Gaussian. The results are given in figure 5.14. The accuracy of the map of Θ_{wp} is reasonable: the standard deviation in Θ_{wp} rarely exceeds 25 per cent of the mean. These maps could be used as the basis of a sensitivity analysis of the WOFOST model with respect to errors in Θ_{wp}.

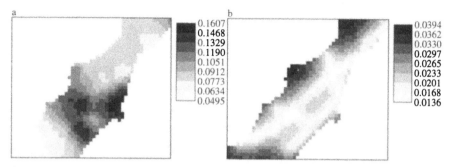

Figure 5.14   Results of the error propagation: (a) block mean and (b) standard deviation of moisture content at wilting point Θ_{wp} (cm^3/cm^3) as obtained with the regression model.

Contribution of error sources

If a sensitivity analysis with WOFOST shows that the errors in Θ_{wp} cause large deviations in the output of WOFOST, then the accuracy of the map of Θ_{wp} needs to be improved. This requires a study to determine which of the error sources makes the largest contribution to the output error. Three error sources can be distinguished here, these are the error in the regression model and the interpolation errors in porosity and moisture content at field capacity. The contribution of each individual error source was determined using the partitioning property discussed in section 4.4. The correlation of the cokriging prediction errors was ignored in the analysis. Figure 5.15 presents results which show that both Θ_{fc} and Φ form the main source of error. Only in the immediate vicinity of the data points is the model a meaningful source of uncertainty, as would be expected because there the cokriging variances of Θ_{fc} and Φ are the smallest.

Thus the main source of error in Θ_{wp} is that associated with the kriging errors of Θ_{fc} and Φ. Improvement of the quality of the map of Θ_{wp} can thus best be done by improving the maps of Θ_{fc} and Φ, by taking more measurements over the study area. The variograms of Θ_{fc} and Φ could be used to assist in optimising sampling (McBratney *et al.*, 1981). This technique would allow one to judge *in advance* how much improvement is to be expected from the extra sampling effort.

Discussion

Several additional points must be addressed with respect to the Allier case study.

Firstly, it is important to note that the correlations of the estimated regression coefficients were found to be negative. This has a neutralising effect on the model error. If zero correlations had been assumed, then this would have caused the model error to become the main source of error, which is a result

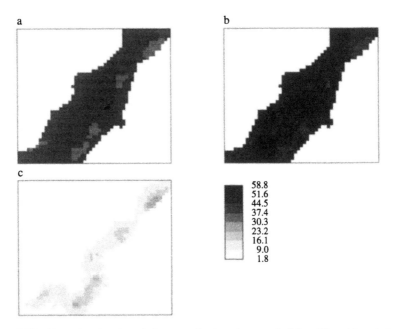

Figure 5.15 Maps showing the relative contributions (per cent) of the different inputs to the variance of calculated moisture content at wilting point: (a) due to moisture content at field capacity, (b) due to porosity, (c) due to the regression model.

that is quite different from the one given in figure 5.15. This clearly shows the risk of carelessly ignoring the presence of non-zero correlations.

Secondly, the change of support that was carried out using block cokriging caused the problem of what to do with the regression residual. Here it was assumed that the residuals average out within a larger block of 50 × 50 m, which is perhaps the most sensible thing to do, but it is still an assumption that may well turn out to be wrong in practice. This would for example occur when the residual contains other soil properties that are not included in the regression, but that do carry information about Θ_{wp} and that are in addition spatially correlated. Again the problem can only be avoided by measuring all properties at the correct support. Unfortunately, this is not feasible here. The purpose of this study was to map Θ_{wp} at a 50 × 50 m grid support for its use in the WOFOST crop growth model, and measuring the soil properties at a support of 50 × 50 m cells is practically impossible.

Thirdly, the Allier case study demonstrated the use of regression models for the mapping of expensive-to-measure attributes, using maps of cheaper-to-measure attributes as input to the regression. This enabled the mapping of the moisture content at wilting point using only 12 direct observations. An interesting alternative to the regression approach would be to use cokriging with the cheap attributes as covariables. The only problem then is that 12 observations of Θ_{wp} are hardly sufficient to model the variogram and cross-variograms of Θ_{wp}. It would be interesting to make a comparison between the methods for situations in which the number of observations do allow for cokriging.

5.4 Selection of suitable soils in the Lacombe agricultural research station using Boolean and continuous classification

A soil survey of a 55 ha part of the agricultural research station at Lacombe in the Black Soil Zone of central Alberta, Canada was carried out in May, 1986 (MacMillan *et al.*, 1987). Data were obtained from soil samples collected at 154 profiles on a 60 × 60 m grid (figure 5.16). Here I use only the results of laboratory analyses of per cent sand (Sand), per cent clay (Clay) and soluble sodium content (Na) of the C horizon (70–80 cm deep). Na was positively skewed and was therefore transformed by computing ln (Na + 1) to give the attribute Lnna. Table 5.3 lists summary statistics for these properties.

Experimental variograms and experimental cross-variograms were computed for all attributes and attribute pairs, and spherical variogram models were then fitted by eye. Table 5.4 presents the fitted parameters. The data from

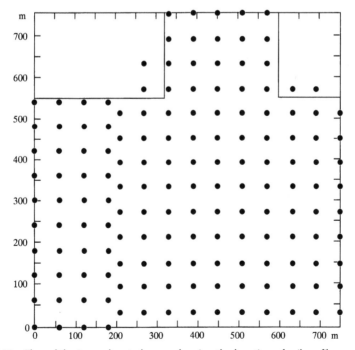

Figure 5.16   Plan of the Lacombe study area showing the location of soil profile observations.

Table 5.3 Summary statistics of the soil data for the Lacombe study.

	Mean	sd	cv	Min	Max	Correlation with Lnna	Sand	Clay
Lnna (ln (meq/100 g))	2.14	1.15	53.9	0.26	4.38	1.00	−0.30	0.36
Sand (per cent)	21.9	20.0	91.5	0.00	65.0	−0.30	1.00	−0.85
Clay (per cent)	41.1	9.3	22.5	21.0	72.0	0.36	−0.85	1.00

the sample points were then interpolated using block cokriging to blocks of 5 × 5 m. Figure 5.17 presents the resulting maps. Maps of the correlations of the interpolation errors are given in figure 5.18. There are small spatial variations in the correlations of the cokriging prediction errors.

Selecting suitable soils using Boolean and continuous classification

Consider the problem of determining the location of 'suitable soils' for agriculture in the surveyed part of the farm. Suitable soils should be non-saline and not too sandy, particularly in the deeper rooting zone. Suppose that sites (5 × 5 m grid cells) will be adjudged 'suitable' if they satisfy the following Boolean criterion:

$$10 \leq \text{Sand} \leq 50 \text{ AND } 15 \leq \text{Clay} \leq 50 \text{ AND } \text{Lnna} \leq 2.5 \qquad (5.5)$$

The problem with the Boolean model (5.5) is that sites are either judged perfectly suitable or not: there is no in-between. Moreover, the Boolean classification rule constitutes a model that is evidently unrealistic. For instance, sites with Clay just below 50 per cent are accepted (provided Sand and Lnna take proper values), whereas sites where Clay is just above 50 per cent are rejected outright. It would be intuitively more satisfactory to replace the Boolean model by one in which suitability continuously varies with clay content (Burrough *et al.*, 1992; Heuvelink and Burrough, 1993).

An alternative to the rigid Boolean classification model is given by *continuous* or *fuzzy* classification (Kandel, 1986; Klir and Folger, 1988; Burrough,

Table 5.4 Parameters of the variograms and cross-variograms for Lnna, Sand and Clay. All variograms are of the spherical type.

	Nugget	Sill	Range (m)
Variogram of Lnna	0.60	1.35	280
Variogram of Sand	250	470	500
Variogram of Clay	55	95	500
Cross-variogram of Lnna and Sand	−1.0	−8.0	550
Cross-variogram of Lnna and Clay	1.4	5.0	600
Cross-variogram of Sand and Clay	−95	−160	550

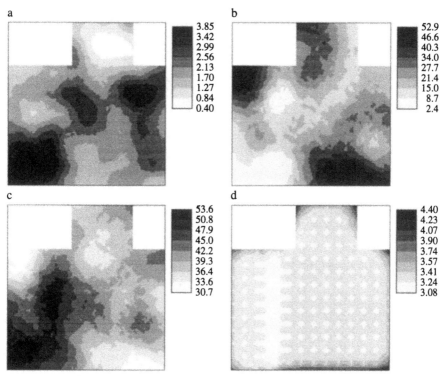

Figure 5.17 Maps of the three soil attributes produced by ordinary block cokriging to a 5 × 5 m cell size: (a) Lnna (ln (meq/100 g)), (b) Sand (per cent), (c) Clay (per cent), (d) cokriging block standard deviation for Clay. (Note all standard deviation maps have the same form but different ranges.)

1989a). Before defining the continuous equivalent of the Boolean model, it is more convenient to reformulate (5.5) in terms of *membership functions*:

JMF^B(Sand,Clay,Lnna)

$$= MIN \{MF^B(Sand,10,50), MF^B(Clay,15,50), MF^B(Lnna, -\infty,2.5)\} \quad (5.6)$$

where JMF^B is the *joint membership function* of the Boolean model. A joint membership function value of 1 means that the site belongs to the class of suitable soils, a value of 0 that it does not. The Boolean membership function MF^B is defined as:

$$MF^B(a,b_1,b_2) = 1 \quad \text{if } b_1 \le a \le b_2 \quad (5.7a)$$

$$= 0 \quad \text{if } a < b_1 \text{ or } a > b_2 \quad (5.7b)$$

where a is the attribute value and b_1 and b_2 are the class boundaries of the Boolean classification model (see figure 5.19a).

To avoid the unrealistic crisp class boundaries of the Boolean membership function MF^B, the continuous membership function MF^C is a function with

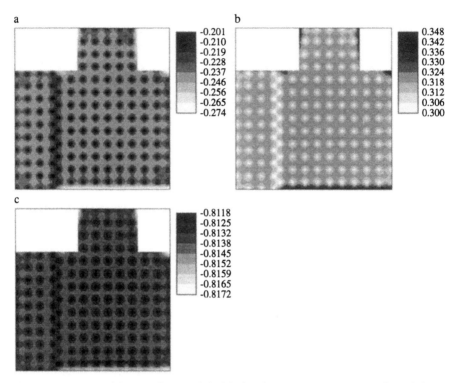

Figure 5.18 Maps of the correlations of the block cokriging errors: (a) Lnna and Sand, (b) Lnna and Clay, (c) Sand and Clay.

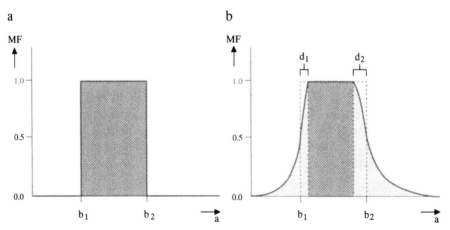

Figure 5.19 The Boolean and continuous membership functions: (a) the conventional Boolean membership function yields a value of 1 inside the class boundary values b_1 and b_2 and 0 elsewhere, (b) a continuous membership function also yields values in between 0 and 1 when in the transition zones around the kernel of the class.

values ranging continuously from 0 to 1 (see figure 5.19b):

$$MF^C(a,b_1,b_2,d_1,d_2) = \frac{1}{1 + \left(\dfrac{a - b_1 - d_1}{d_1}\right)^2} \quad \text{if } a < b_1 + d_1 \tag{5.8a}$$

$$= 1 \quad \text{if } b_1 + d_1 \le a \le b_2 - d_2 \tag{5.8b}$$

$$= \frac{1}{1 + \left(\dfrac{a - b_2 + d_2}{d_2}\right)^2} \quad \text{if } a > b_2 - d_2 \tag{5.8c}$$

where the parameters d_1 and d_2 determine the width of the transition zones around the kernel of the class. These transition zones are chosen such that the value of the continuous membership function equals one half when at the Boolean thresholds b_1 and b_2. Note that if the parameters d_1 and d_2 are zero, (5.8) yields the Boolean membership function (5.7).

For the selection of suitable soils in the Lacombe study area, the transition zone parameters d_1 and d_2 were taken as 5 per cent for Clay and Sand, and 0.3 ln (meq/100 g) for Lnna. These values were considered realistic for this application (Heuvelink and Burrough, 1993). The continuous joint membership function was therefore chosen as:

$$JMF^C(\text{Sand,Clay,Lnna}) = MIN\{MF^C(\text{Sand},10,50,5,5),$$

$$MF^C(\text{Clay},15,50,5,5), \ MF^C(\text{Lnna}, -\infty,2.5,0.3,0.3)\} \tag{5.9}$$

To sum up, continuous classification transforms the data to a continuous scale ranging from 0 to 1, where the value of the membership function gives the *degree* to which the site belongs to the class. Continuous classification is thus able to deal with the problem of unrealistic sharp class boundaries in a simple and intuitive way: the only problem is to select the values of the transition zone parameters d_1 and d_2.

Error propagation with Boolean and continuous classification

In order to understand how errors in the data affect the results of Boolean and continuous classification, figure 5.20 graphically illustrates several possible situations that can arise when an error-contaminated attribute A with probability density f_A, mean b and standard deviation σ is to be placed in a class whose boundaries are b_1 and b_2. Note that figure 5.20 represents the general situation and so it is not restricted to the Lacombe case study. It is included here because it nicely shows how error propagates through Boolean and continuous classification and also because it will make the interpretation of the results for the Lacombe case study easier. These results will be presented later on in this section.

Figure 5.20a shows the standard Boolean case when there is no error, so $\sigma = 0$. The attribute A is in effect deterministic so the individual observation either falls entirely within the class boundaries (right-hand bar) or it falls

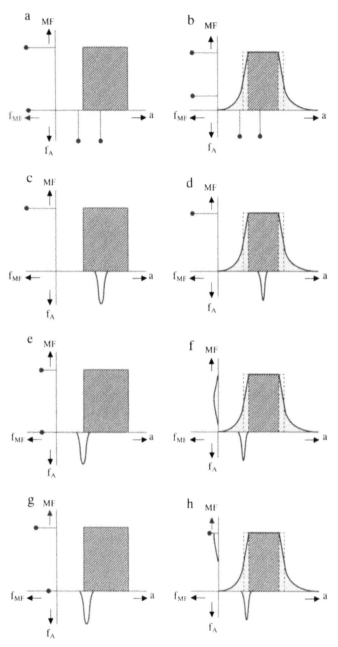

Figure 5.20 Examples of error propagation through Boolean and continuous classification: the output error is zero when the input is exactly known (a and b) or when the input distribution is well within the class limits (c and d). Error in the output occurs when the input distribution straddles the Boolean class boundaries (e and g) or the continuous transition zones (f and h).

outside (left-hand bar). The corresponding values of the membership function are 1 or 0 respectively. Because σ is zero the membership value MF(A) is also error-free. Figure 5.20b shows the same situation where the individual observation is now classified by a continuous membership function. The right-hand observation is within the class kernel so MF(A) = 1. The left-hand observation is outside the kernel and the Boolean boundaries so it returns an MF(A) < 0.5.

Now consider what happens when σ is non-zero. In figure 5.20c, the probability density f_A of A takes the characteristic bell-shaped form and lies well within the limits of the Boolean class. The probability that A really falls outside the class limits is vanishingly small so for all practical purposes MF(A) = 1. The same situation occurs with the continuous membership function when f_A falls within the class kernel (figure 5.20d). Equally clear-cut results follow when f_A lies well outside the class limits. So in these cases the error in the observation does not cause an error in the classification result. This is in itself a result that is worth noting.

The results are less than clear-cut when f_A straddles the Boolean class boundaries or the continuous transition zones. Figures 5.20e,f show the situation when the mean of A equals b_1. In the Boolean case (figure 5.20e) the distribution of MF(A) becomes a discrete distribution with two possible values, 0 and 1. In this case the chances are equal that A falls within or outside the class boundary so the probability of obtaining each value is 0.5. In the continuous case shown in figure 5.20f, the mean of A equals b_1 and the distribution band is narrower than the transition zone. In this case MF(A) is continuously distributed just as A is. The mean of MF(A) is about 0.5 and because the membership function varies steeply at b_1, the standard deviation of MF(A) is much larger than for A, though it is still substantially smaller than the Boolean standard deviation in figure 5.20e.

Figures 5.20g,h show the situation when f_A mainly lies inside the Boolean limit b_1, but still with considerable overlap. In the Boolean case (figure 5.20g) the effect is to unbalance the probabilities of returning a membership value of 0 or 1; the probability of returning a value of 1 is increased. In the continuous case f_A runs into the class kernel so the resulting distribution is a mixed distribution, i.e. the combination of a continuous and a discrete distribution. There is a definite chance that the membership value will be exactly 1 in some cases, though values less than 1 can also occur. Although not shown, the situation where f_A is broader than the transition zone will generally yield mixed distributions of MF(A) with peaks at 0 and/or 1.

Results for the Lacombe suitability analysis

The Monte Carlo method was applied to the Boolean and continuous classification models (5.6) and (5.9). The number of runs was taken as $N = 2000$. It was assumed that the attributes and the associated errors are joint normally distributed with parameters as obtained from block cokriging.

Figure 5.21 presents the results of the Boolean and continuous models, as applied to the interpolated surfaces of figures 5.17a,b,c. Note that the presence of interpolation errors is not yet under consideration. Whereas the Boolean model divides the area into two classes (0 or 1), continuous classification clearly shows the gradation of suitability. Given the continuous nature of the variation of the soil properties in the area, this result seems entirely reasonable. Continuous classification provides the user with more information because it can distinguish smaller differences between cells, whereas with Boolean classification a cell can only be suitable or not suitable. Note also that figure 5.21b contains figure 5.21a because whenever $JMF^C < 0.5$ then $JMF^B = 0$, and when $JMF^C > 0.5$ then $JMF^B = 1$.

Now consider the results of the Monte Carlo simulations (figure 5.22). First observe that figures 5.22a and 5.22b are very similar, which means that the expected Boolean JMF^B value is almost the same as the expected continuous JMF^C value. This is not surprising because the continuous transition zones were placed symmetrically over the Boolean boundaries, which ensures that neither of the models will *on average* yield higher membership values.

Figures 5.22a,b also reflect the patterns of figure 5.21, although the average map values of figure 5.21 are noticeably larger. This is caused by the Minimise operation, because on average the *minimum* of the attribute values is smaller when they are perturbed by error. This is a clear example of what I noted in Chapter 4, namely that the mean output need not equal the output of the mean input. An important consequence is that there is probably less land satisfying (5.5) than is suggested by figure 5.21. The Boolean map of figure 5.22a no longer contains only the extreme values 0 and 1, but also values between 0 and 1. These values should *not* be interpreted as continuous membership values, however: they are the *average* Boolean JMF^B values, which are just the proportion of ones that occurred during the Monte Carlo

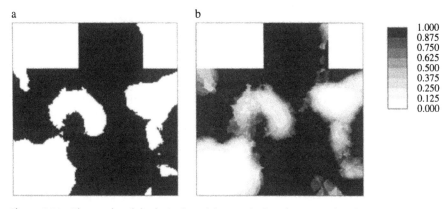

Figure 5.21 The results of the logical model as applied to the interpolated surfaces: (a) Boolean classification, (b) continuous classification.

simulation. Figure 5.22a can thus also be interpreted as a map giving for each cell an estimate of the *probability* that it is accepted. Similarly, figure 5.22b is a map of the average JMF^C values obtained by simulation.

Figures 5.22c,d show the standard deviations of the simulated surfaces. The overall standard deviations are quite large, especially for the Boolean model. For the Boolean model there are large areas where the standard deviation is close to one half, which is the theoretical maximum. These large standard deviations demonstrate that decisions based on figure 5.21 may frequently turn out to be wrong in practice. The standard deviations are the largest in locations where the attribute values are near the Boolean class boundaries or within the continuous transition zones. Grid cells that are clearly within or clearly excluded from a class or classes are selected or rejected with a low level of uncertainty. Comparison of figures 5.22c,d also shows that for continuous classification the levels of uncertainty are substantially lower than for Boolean classification. This agrees with the discussion of figure 5.20.

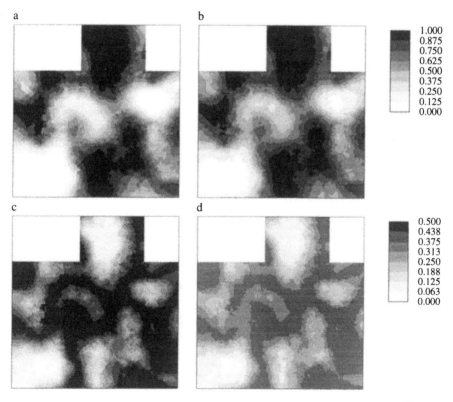

Figure 5.22   The results of Monte Carlo simulation using 2000 samples: (a) mean JMF^B value for the Boolean classification model, (b) mean JMF^C value for the continuous classification model, (c) standard deviation of JMF^B, (d) standard deviation of JMF^C.

Figure 5.23 is figure 5.22a masked by figure 5.21a. It shows the probability of an *incorrect decision* when the strict Boolean model is applied to the interpolated surfaces of figure 5.17. Figure 5.23a shows that the strict Boolean model selects many cells even though the probability of a cell actually being suitable can be much less than 1. Figure 5.23b shows that some cells are rejected even though their probability of being suitable is almost equal to 0.5. It is remarkable that some of the rejected cells have a larger probability of being suitable than some of the accepted cells. Figure 5.23 thus shows that with the Boolean model there is quite a considerable risk of making incorrect decisions when they are solely based on the interpolated surfaces. Note also that such incorrect decisions mainly take place where the value of JMF^C is nearly one half (see figure 5.21b).

Comparison of Boolean and continuous classification

When Boolean classes are used for classification of environmental data the results may be unsatisfactory, because many environmental problems cannot realistically be modelled using the rigid Boolean classification rules. Continuous classification provides a solution to this problem because it relaxes the discrete class membership values to continuous class membership values, thus allowing users to define flexible class membership functions that match practical experience (Burrough, 1989a; Burrough *et al.*, 1992). This case study shows that continuous classification is also advantageous because it is less sensitive to errors in the data. Theoretical arguments and the results of the case study show that standard deviations of the membership values are substantially smaller for continuous classification than for Boolean classification. Recently, these findings were confirmed by a case study on irrigated riceland in the Philippines (Dobermann and Oberthür, 1997).

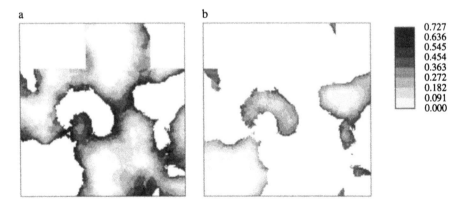

Figure 5.23 Probability of incorrect decisions based on Boolean classification of the interpolated surfaces: (a) the probability that a selected cell should be rejected, (b) the probability that a rejected cell should be selected.

CHAPTER SIX

Error propagation with global GIS operations: the use of multidimensional simulation

The common feature of global GIS operations is that their result at some location may depend on input values at locations in the study area that are remote from the location of interest. In other words, global operations include far-reaching spatial interactions. Typical examples of global operations are Spread, Viewshed, Stream and Pathway (Burrough, 1986; Tomlin, 1990; Burrough and McDonnell, 1998). This type of operation is routinely applied to digital elevation models and cost surfaces, such as used to determine the optimal trajectory of a highway in a landscape (Tomlin, 1983), or for automated drainage network analysis (Kovar and Nachtnebel, 1993; Van Deursen, 1995).

Distributed dynamic models

An important class of global operations is formed by distributed dynamic models, such as groundwater and hydrological catchment models. These models usually work with time steps and although each time step only involves neighbourhood operations, global influences will occur over a longer period of time. For instance, the flow at the outlet of a catchment is affected by the infiltration capacity of the soil throughout the catchment, not only in its immediate vicinity (De Roo, 1993).

The propagation of errors in distributed dynamic models can be analysed using the Taylor method. This has proven useful in groundwater modelling (Dettinger and Wilson, 1981; Wagner and Gorelick, 1987; Jones, 1989) and elsewhere as well (Scavia *et al.*, 1981; Kuczera, 1988; Loague *et al.*, 1989). However, application of the Taylor method requires a case-specific approach and an additional problem is to judge whether the approximations invoked by the Taylor method are always acceptable (Kuczera, 1988). Because this research aims to present a methodology that is valid for a broad class of global operations, I only consider the Monte Carlo method here.

6.1 Monte Carlo method for global operations

In chapter 4 the flexibility of the Monte Carlo method was mentioned as one of its main advantages. This flexibility also becomes apparent when we consider the application of the method to a global operation $g(\cdot)$:

1 Repeat N times:
    (a) Generate a set of input realisations $a_i(\cdot)$, $i = 1, \ldots, m$.
    (b) For this set of input realisations $a_i(\cdot)$, compute and store the output
        $u(\cdot) = g(a_1(\cdot), \ldots, a_m(\cdot))$.
2 Compute and store sample statistics from the N outputs $u(\cdot)$.

The procedure is thus very much the same as the one given in section 4.2.4. The only meaningful difference is that step 1(a) requires the simultaneous generation of realisations from the random fields $A_i(\cdot)$, sometimes also referred to as multidimensional simulation (Brooker, 1985; Sharp and Aroian, 1985; Heuvelink, 1992). This is more difficult than sequentially simulating from the random variables $A_i(x)$ for all $x \in D$, because of the presence of spatial correlation.

Once the problem of efficiently generating realisations of correlated random fields is resolved, the application of the Monte Carlo method to error propagation with global operations is straightforward. A recent example of applying the method to the Viewshed operation is given by Fisher (1992). There are also numerous applications to distributed hydrologic models (Smith and Freeze, 1979; Smith and Hebbert, 1979; Peck et al., 1988; Jones, 1989; De Roo et al., 1992; Binley et al., 1997). The main problem with the method is the computing time required. A single run of a groundwater or erosion model may easily take an hour on a mini-computer, and the entire error analysis will take at least N times as much.

Application of the Monte Carlo method to distributed dynamic models is thus fairly easy, provided there are sufficient computing resources and efficient ways to generate realisations of spatially correlated random fields. In the rest of this chapter I will concentrate on the latter problem, to begin with the univariate situation ($m = 1$). Note also that the discussion below is restricted to the Gaussian situation.

6.2 Stochastic simulation of the input random field

Generating a realisation $a(\cdot)$ of $A(\cdot)$ is commonly referred to as *conditional simulation*. This is because $A(\cdot)$ is the random field $Z(\cdot)$ conditioned to point observations $z(x_i)$, as discussed in section 2.3.2. Simulating from $Z(\cdot)$ thus, without conditioning the realisation to the point observations, is known as *unconditional simulation*.

In practice, we will only want to simulate $A(\cdot)$ at a finite number of locations $x_j, j = 1, \ldots, J$. In most cases this will be a grid of nodes. Since the $A(x_j)$ constitute a vector of correlated random variables, in principle realisations can be generated using the methods of multivariate simulation (Johnson, 1987). One possible method uses the Cholesky decomposition already mentioned in section 4.2.4. However, the large dimension J (for example, a 200×200 grid yields $J = 40\,000$) makes this approach of limited use for practical purposes (Davis, 1987a). Some suggestions have been made to modify the method to make it suitable for larger grids (Alabert, 1987; Davis, 1987b), but these then suffer from approximation errors and involve more complicated algorithms (Dowd, 1992).

When $A(\cdot)$ is the result of a kriging interpolation, then a two-step simulation approach is often used. Firstly, an unconditional realisation $z(\cdot)$ of $Z(\cdot)$ is generated. Secondly, the realisation $z(\cdot)$ is conditioned to the observations using simple kriging (Journel and Huijbregts, 1978; Delhomme, 1979; Cressie, 1991). The two-step approach is attractive because unconditional simulation is relatively easy: it can exploit the stationarity of $Z(\cdot)$. There are various methods of unconditional simulation, the most important of which are the turning bands method (Matheron, 1973; Journel and Huijbregts, 1978; Brooker, 1985; Christakos, 1987), spectral methods (Mejia and Rodriguez-Iturbe, 1974; Borgman et al., 1984; Bellin et al., 1992) and the moving average method (Journel and Huijbregts, 1978; Matérn, 1986).

Less than a decade ago, an alternative simulation method was proposed, known as sequential Gaussian simulation (Journel and Alabert, 1989; 1990). One of the advantages of this method is that the extension from unconditional to conditional simulation is very easy. Another advantage is that the change of support can easily be accounted for. Using this method all grid nodes x_j are visited randomly, each time computing and drawing from the conditional distribution of $A(x_j)$. Simple kriging is used to condition $A(x_j)$ to the nodes already generated. Further advantages of the sequential Gaussian simulation algorithm are that it can handle any type of covariance structure and that approximation errors are usually small. The method has also been extended to the non-Gaussian situation, by a method known as sequential indicator simulation (Journel and Alabert, 1989; Gómez-Hernández and Srivastava, 1990; Deutsch and Journel, 1992). Recently, much of the research effort of the geostatistics community has been spent on spatial stochastic simulation (Journel, 1996; Baafi and Schofield, 1997).

Extension to multivariate spatial simulation

When there are multiple correlated input random fields $A_i(\cdot)$, then these must be simulated simultaneously. There are at least two viable approaches for this.

The first is the technique of joint sequential simulation of multi-Gaussian random fields (Gómez-Hernández and Journel, 1993). This is a natural extension of sequential Gaussian simulation, the main difference being that each conditioning step is made by simple cokriging, instead of simple kriging.

The second approach is by employing the linear model of coregionalisation of multivariate geostatistics (Journel and Huijbregts, 1978; Goovaerts, 1992; Goulard and Voltz, 1992). Here the idea is to write each of the $A_i(\cdot)$ as a weighted sum of independent underlying random fields. These underlying random fields can then separately be simulated, using one of the univariate methods described above.

6.3 An iterative method for simulating autoregressive random fields

All simulation methods described so far assume that either the covariance function $R(\cdot,\cdot)$ of the input attribute $A(\cdot)$ is explicitly known or, when a two-step simulation is employed, that the covariance function $C(\cdot)$ or variogram $\gamma(\cdot)$ of $Z(\cdot)$ is known. The requirement that $\gamma(\cdot)$ is known is unlikely to cause problems in geostatistical applications, because the structural analysis, i.e. the construction of the variogram from point observations, is 'the first and indispensable step in any geostatistical study' (Journel and Huijbregts, 1978, p. 12). However, there are also situations in which the correlation structure of $Z(\cdot)$ is only implicitly known. In such situations it may be impractical to derive the covariance function of $Z(\cdot)$ first and then use one of the methods described above to simulate $Z(\cdot)$ at grid nodes. An important class of such random fields is those defined by spatial autoregressive models (Bennett, 1979; Griffith, 1988; Upton and Fingleton, 1989; Cressie, 1991).

Random fields that are defined by a spatial autoregressive model are not fundamentally different from the 'geostatistical' random fields we have considered so far. However, they differ from the geostatistical models in two meaningful ways. The first is that the domain D is usually restricted to a lattice of locations, such as a grid of nodes or an irregular configuration of spatial locations (Cressie, 1991). The second is that they presuppose that the spatial correlation structure of the random field satisfies a conceptual statistical model, such as a model of linear nearest neighbour dependency:

$$Z(x) = \mu + \sum_i q_i(Z(x_i) - \mu) + rW(x) \quad x, x_i \in D \tag{6.1}$$

where the q_i and r are parameters, μ is the general mean of $Z(\cdot)$, the x_i are neighbours of x and where $W(\cdot)$ is a zero-mean, spatially independent noise field.

Spatial autoregressive models are frequently employed in geographical research (see for example Griffith (1988) or Cressie (1991) and the references therein). Moreover, much of the error modelling in GIS is done by means of spatial autoregressive models (Goodchild and Gopal, 1989; Fisher, 1992; Goodchild et al., 1992; Hunter and Goodchild, 1997). Therefore it is important that attention is paid here to the simulation of autoregressive random fields, i.e. of random fields whose spatial autocovariance is not explicitly given but embedded in a conceptual statistical model such as (6.1).

In this section I present an iterative method for simulating an autoregressive random field. I will do so by means of a simple first order autoregressive model, but the method has general applicability. The method aims to overcome the computational overhead associated with the traditional inversion method. A similar approach has more recently and independently been proposed by Zhang and Yang (1996).

Definition of the first order autoregressive random field

The simplest, two-dimensional, non-degenerate form of the spatial autoregressive model considers a random field $Z[\cdot,\cdot]$ defined on a regular grid:

$$Z[k,l] = q(Z[k-1,l] + Z[k+1,l] + Z[k,l-1] + Z[k,l+1]) + rW[k,l]$$

$$(6.2)$$

where $1 \leq k \leq K$, $1 \leq l \leq L$, $0 \leq q < 0.25$ and $r \in \mathbb{R}$. The zero-mean noise field $W[\cdot,\cdot]$ is assumed normally distributed, with covariance $\mathrm{cov}(W[k,l], W[k',l']) = 1$ if $k = k'$ and $l = l'$, $\mathrm{cov}(W[k,l],W[k',l']) = 0$ if $k \neq k'$ or $l \neq l'$. Note that a somewhat different notation is used here to underline the fact that $Z[\cdot,\cdot]$ and $W[\cdot,\cdot]$ are defined on a two-dimensional regular grid. Note also that for convenience, but without loss of generality, it is assumed that $\mu = 0$.

The random field $Z[\cdot,\cdot]$ defined by (6.2) is known as the first order simultaneously specified spatial Gaussian model (Cressie, 1991), abbreviated here to first order spatial autoregressive model. It is also referred to as the first order nearest neighbour model (King and Smith, 1988; Heuvelink, 1992).

Here I will be concerned with the model (6.2) only, with boundary conditions:

$$Z[k,0] = Z[k,L+1] = Z[0,l] = Z[K+1,l] = 0 \qquad (6.3)$$

However, the results presented here can easily be extended to the general type of spatial autoregressive models with arbitrary boundary conditions.

Whittle (1954) showed that when boundary effects are negligible, $Z[\cdot,\cdot]$ is the discretisation of a stationary Gaussian continuous-space random field with zero mean and a Bessel type of correlation function. More recently, King and Smith (1988) showed that the associated variogram $\gamma(\cdot)$ is given by:

$$\gamma(h) = \frac{2r^2}{\pi(1 - (4q)^2)} E(4q)(1 - \kappa h K_1(\kappa h)) \qquad (6.4)$$

where $\kappa = \sqrt{(1/q - 4)/\Delta}$, Δ is the grid mesh, $K_1(\cdot)$ is the modified Bessel function of the first order and the second kind and $E(\cdot)$ is the complete elliptic integral of the second kind. The Bessel variogram (6.4) can be rapidly computed using standard library routines as in Press *et al.* (1992). It is given in figure 6.1. The shape of $\gamma(\cdot)$ resembles that of the well-known Gaussian variogram, with zero nugget and an effective range of $4\Delta/\kappa$.

Inversion method for autoregressive spatial simulation

In order to describe the inversion method for autoregressive spatial simulation, define $K \cdot L$-dimensional stochastic vectors $\bar{Z}$ and $\bar{W}$ with $\bar{Z}_{(k-1)L+l} = Z[k,l]$ and $\bar{W}_{(k-1)L+l} = W[k,l]$, $k = 1, \ldots, K$, $l = 1, \ldots, L$, and construct a matrix $Q \in \mathbb{R}^{K \cdot L} \times \mathbb{R}^{K \cdot L}$ such that $Q_{ij} = q$ if $\bar{Z}_i$ and $\bar{Z}_j$ are 'neighbours' and $Q_{ij} = 0$ if they are not. In this way, (6.2) can be rewritten as:

$$\bar{Z} = Q\bar{Z} + r\bar{W} \tag{6.5}$$

From which follows:

$$\bar{Z} = r(I - Q)^{-1}\bar{W} \tag{6.6}$$

Now since realisations $\bar{w}$ of $\bar{W}$ can easily be generated, the problem is transformed into one of inverting the $K \cdot L \times K \cdot L$ matrix $(I - Q)$. Once $(I - Q)^{-1}$ is obtained, many realisations $\bar{z}$ of $\bar{Z}$ can be generated by applying (6.6) to realisations $\bar{w}$ of $\bar{W}$. In fact, this approach has much resemblance to the Cholesky decomposition method (Davis, 1987a), because $r(I - Q)^{-1}$ is the square root of the covariance matrix of $\bar{Z}$.

It will not be surprising that the numerical load associated with the inversion quickly becomes too demanding when the size of the grid increases, i.e. when K and L are large (Haining *et al.*, 1983; Goodchild *et al.*, 1992). Moreover, the matrix $(I - Q)$ becomes ill-conditioned when q is close to 0.25,

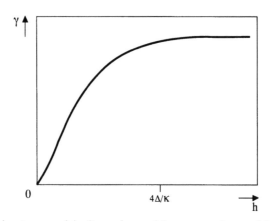

Figure 6.1 Bessel variogram of the first order spatial autoregressive model.

causing errors in the numerical inversion procedure (Baker, 1984). Sharp and Aroian (1985) propose to overcome these problems by starting from a non-symmetric unilateral nearest neighbour model, but a serious disadvantage of their approach is that the simulated field is no longer stationary and that the relationship with a continuous-space random field with a Bessel variogram can no longer be established. King and Smith (1988) used the known shape of $(I - Q)$ to calculate its inverse analytically. However, this still leaves numerical problems and the results cannot easily be extended to the case of general boundary conditions or models more complicated than (6.2).

Description of the iterative simulation method

The following simple iterative method is proposed to generate a realisation $\bar{z}$ of $\bar{Z}$:

1 Generate a realisation $\bar{w}$ of $\bar{W}$.

2 Take initial $\bar{z}(0) = r\bar{w}$.

3 Given $\bar{z}(k)$, calculate $\bar{z}(k + 1) = Q\bar{z}(k) + r\bar{w}$.

4 If $\bar{z}(k + 1) = \bar{z}(k)$ then ready, else increase k and repeat step 3.

Clearly, once the condition in step 4 is satisfied, the resulting $\bar{z}(k)$ is the unique solution to (6.5) for the given $\bar{w}$. It remains to prove that the iteration converges. To prove this, it is sufficient to show that the k-th power of Q converges to the zero matrix:

$$\lim_{k \to \infty} Q^k = 0 \qquad (6.7)$$

To show that (6.7) holds, construct a $(K{\cdot}L + 1) \times (K{\cdot}L + 1)$ matrix P as follows:

$$P = \begin{bmatrix} & & p_1 \\ & & p_2 \\ & Q & \vdots \\ & & p_{K \cdot L} \\ 0 \cdots 0 & & 1 \end{bmatrix} \qquad (6.8)$$

where $p_i = 1 - Q_{i1} - Q_{i2} - \ldots - Q_{iK \cdot L}$ for all $i = 1, \ldots, K{\cdot}L$. Note that all $p_i > 0$ since $q < 0.25$. Now since P is the transition matrix of a Markov chain whose first $K \cdot L$ states are transient and whose last state is absorbing, the proof is completed by applying a limit theorem from the theory of Markov chains (Cinlar, 1975, theorem 5.3.2), which states that:

$$\lim_{k \to \infty} (P^k)_{ij} = 0 \tag{6.9}$$

for $1 \leq i \leq K{\cdot}L + 1$, $1 \leq j \leq K{\cdot}L$. It can easily be verified that (6.9) implies (6.7).

That (6.9) holds can, in the terminology of the theory of Markov chains, be explained as follows. At each step with probability p_i one moves from the transient state i to the absorbing state $K \cdot L + 1$. Since p_i is greater than zero for all i one is bound to arrive eventually at the absorbing state, where one stays permanently. Thus the limiting probability is 0 for all transient states and 1 for the absorbing state. Presumably, when the p_i are close to zero many steps are needed before the limiting probability is reached. This, in turn, implies that the iteration will take many more steps when q is close to the value of 0.25.

The obvious way to implement the iteration is repeatedly to move through the grid on a cell-by-cell basis, at each cell applying (6.2) to the k-th $\bar{z}$ to yield the $(k + 1)$-th $\bar{z}$. Thus each iteration step involves $K{\cdot}L$ times evaluating (6.2), so that the method is linear in the number of grid cells. The iteration is stopped once for all cells the difference between the k-th and the $(k + 1)$-th $\bar{z}$ is less than some critical value δ.

The Jacobi type of iteration as described above can be improved by using a modified Gauss–Seidel iteration, in which newly calculated values are directly being used by the method (Schuit, 1989). This leads to a reduction of both the number of iterations and the storage capacity required. It is not difficult to show that also other changes to the model, such as direction dependent q's, inclusion of higher order neighbours, multidimensional models or reflecting boundary conditions do not affect the validity of the iterative method.

6.4   Numerical experiments with iterative autoregressive simulation

Simulations of a 200×200 grid with $\mu = 100$ and $\Delta = 1$ were carried out for ten different values of q, ranging from $q = 0.15$ to $q = 0.2495$. Realisations of the initial noise field $W[\,\cdot\,,\cdot\,]$ were taken from a standard normal distribution, using the Ran1 algorithm as a pseudo random number generator (Press *et al.*, 1992). The parameter r was chosen such that the variance of $Z[\,\cdot\,,\cdot\,]$, which equals the sill of the variogram (6.4), was 100. The critical value δ was set to 0.001. It was assumed that this value for δ was small enough to guarantee that the iteration had converged; this was confirmed by comparisons between maps that were generated with $\delta = 0.001$ and $\delta = 0.0001$, where no noticeable differences could be demonstrated. An example showing several intermediate steps of the iteration is given in figure 6.2.

Table 6.1 reports the results for different values of q. The simulations were carried out in 1992 on a 486-DX 33 MHz personal computer. Some of the

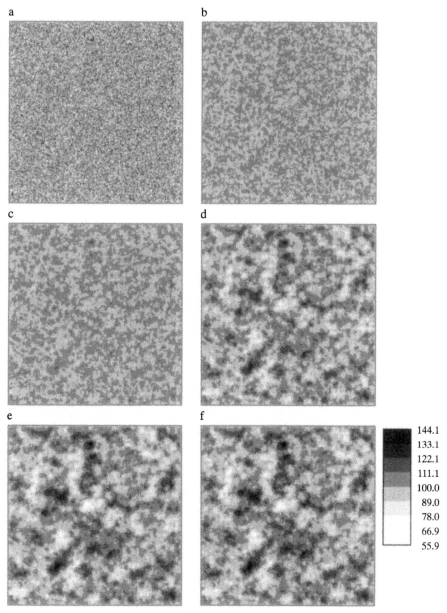

Figure 6.2   Intermediate results for iterative autoregressive simulation with $q = 0.2470$: (a) initial noise field, (b) after 2 iterations, (c) after 5 iterations, (d) after 30 iterations, (e) after 100 iterations, (f) final result with δ set to 0.001 (after 218 iterations).

Table 6.1   Numerical results for the simulation of a 200 × 200 grid.

q	$4/\kappa$	Number of iterations	Computing time (s)
0.1500	2.4	11	3
0.2000	4.0	21	6
0.2300	6.8	46	13
0.2400	9.8	83	24
0.2450	14.0	146	42
0.2470	18.1	229	66
0.2480	22.3	315	90
0.2490	31.6	534	154
0.2493	37.7	637	183
0.2495	44.7	929	269

accompanying example realisations are given in figure 6.3. Note the increase of spatial correlation with an increase of q. The results are quite satisfactory. The computing time required is low for small q and moderate for q close to 0.25. The attractiveness of the method is evidently clear when one realises that in this case the inversion method would require the inversion of a 40 000 × 40 000 matrix. In spite of the inherent simplicity of the matrix Q, this is still an arduous task (Goodchild *et al.*, 1992).

Discussion of results

The results clearly show that the convergence rate of the method is sensitive to the value of q. Judging from the figures in table 6.1, it appears that the number of iterations required doubles when $(1 - 4q)$ is halved. It is not difficult to show that this result is in agreement with the limiting behaviour of the Markov chain discussed in the previous section.

Because of the iterative nature of the method, a suitable criterion is needed to decide when the iteration has sufficiently converged. In this study the convergence criterion was safely chosen such that adding more iterations did not noticeably alter the result. However, preferably one would derive the criterion directly from maximum error bounds allowed on the obtained result. Therefore more research is needed on the convergence properties of the method. The theory of Markov chains seems to form an adequate basis for this.

The iterative method as described in this and the previous section provides an attractive method for the simulation of autoregressive spatial processes. Not only are the computing time and storage capacity required by the method orders of magnitude smaller than for the inversion method, but the generality and simplicity of the method facilitate a straightforward implementation of almost any type of model. The numerical complexity of the method is

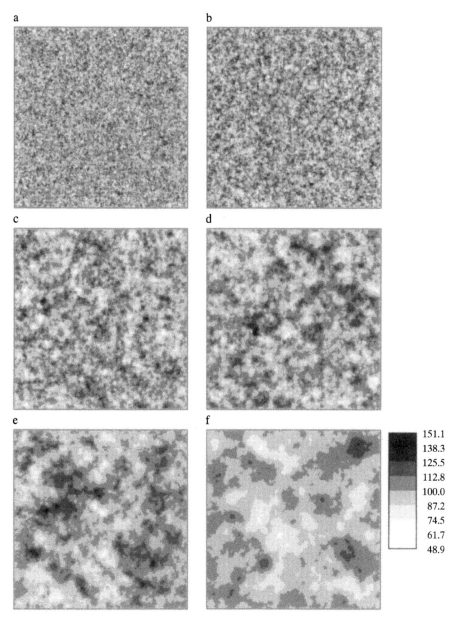

Figure 6.3   Example realisations for different values of the spatial autocorrelation parameter q: (a) $q = 0.1500$, (b) $q = 0.2000$, (c) $q = 0.2400$, (d) $q = 0.2450$, (e) $q = 0.2480$, (f) $q = 0.2495$.

proportional to the number of grid cells, which makes it especially attractive for the simulation of larger grids. A disadvantage of the method is that for each realisation the iteration needs to be repeated. This makes the method less attractive for generating many realisations of a field of a relatively small size.

CHAPTER SEVEN

Implementation of error propagation techniques in GIS

In previous chapters I have paid much attention to the theory and application of error propagation in enviromental modelling with GIS. Multiple techniques can be used to analyse how errors propagate through GIS operations, but none of these techniques has as yet been fully incorporated in any professional GIS package. Thus an important question is how error propagation techniques can best be implemented in a GIS.

This chapter discusses some of the problems that need to be addressed when error propagation functionality is to be added to a GIS. It also describes an error propagation software tool (named ADAM) that was developed in this research and that was used to run the applications in chapter 5. The purpose of this chapter is not to elaborate on the many features of ADAM: these are described in Wesseling and Heuvelink (1991); (1993a).

7.1 Starting points for adding error propagation functionality to an existing GIS

There are various ways in which error propagation functionality can be added to a GIS. In this research, the following starting points were formulated from which the error propagation tool should be developed:

1 *The error propagation tool should not replace the GIS*
  The error propagation tool should make as much use as possible of the existing storage and analytical capabilities of the GIS. It should not define

its own database structure, and neither should it perform any calculations that can be done by the GIS. In other words, the tool should *add* functionality to the GIS, not *replace* functionality.

2 *The error propagation tool should be flexible*
Firstly, the tool should be flexible in the sense that it can handle a variety of GIS operations. Secondly, it should also be flexible in the sense that it can easily be linked to various types of GISs. It should not be a tool that can only work with one specific GIS or data structure. It should work with both vector and raster data structures.

3 *The user interfacing should be efficient and allow for exploratory use of the tool*
Much has to be specified to perform an error analysis. The user must define the GIS operation or computational model, the inputs and their statistical properties. In many cases it will not be sufficient to provide only the means and standard deviations of the inputs. Many analyses will also require the distribution type and/or auto- and cross-correlations. The error propagation tool must find an efficient way of obtaining this information from the user. Moreover, it should also enable the user to examine, in an exploratory way, how output error varies under changes in input error parameters.

4 *The tool should employ efficient computational routines to minimise the numerical load*
Given the fact that some of the error propagation techniques can be very computationally demanding, there is much to be gained by choosing efficient numerical algorithms and by including common code optimisation techniques (Aho *et al.*, 1986). In this way important savings in computing time can be made.

7.2   The ADAM error propagation software tool

The four points above have been considered in the development of ADAM, a tool for tracing error propagation in quantitative spatial modelling with GIS (Wesseling and Heuvelink, 1993a).

ADAM is a tool that does not perform any error propagation calculations itself. It merely functions as a compiler, translating a user's error propagation problem into standard GIS operations. The user feeds the problem to ADAM by means of the model description language (see below). ADAM then analyses the problem and after an appropriate error propagation technique has been selected, it generates a string of standard GIS operations that embody the error propagation analysis. This string of operations must then be carried out by the GIS to obtain the results of the error propagation analysis.

ADAM identifies the input attributes to the error propagation by symbolic names. These symbolic names refer to data layers in the GIS, but ADAM makes no assumptions about the data structure of these layers. In this way it achieves the required data structure flexibility. Since ADAM leaves the actual calculations to the GIS, it also satisfies the principle that it should only add functionality to the GIS.

The model description language (MDL)

ADAM uses the so-called model description language (MDL) as a user interface. Using a 'programming language' as an interface is appropriate here because the various language constructions that MDL contains allow one to define almost any type of error propagation problem. Moreover, the syntax of MDL offers great flexibility in its use.

There are four different ways of declaring the inputs to a GIS operation in MDL. This depends on whether the input is deterministic or stochastic, and on whether it is spatial or non-spatial. Examples of the four types of input declaration are:

Deterministic and non-spatial:	A: 48.650;
Deterministic and spatial:	B: "b.map";
Stochastic and non-spatial:	C: random(normal, $-1.076,0.056$);
Stochastic and spatial:	D: random(unknown,"d.est","d.sdv");

Spatial attributes are identified by symbolic names that refer to GIS data layers. Stochastic inputs are declared by a distribution type followed by distribution parameters, which are usually the mean and standard deviation (Wesseling and Heuvelink, 1993a).

MDL gives considerable flexibility in specifying stochastic parameters. For instance, spatial attributes with non-spatial standard deviations or constant coefficient of variation can be defined as:

E: random(normal,"e.est",0.80);
F: random(normal,"f.est",0.25*"f.est");

Auto- and cross-correlation are identified in MDL by the symbol 'r'. As an example, consider the declaration:

r(G,H): exponential($-0.68,38.0$);

This means that the cross-correlation of G and H for two locations a distance h apart ($h > 0$) is given by $-0.68 \cdot \exp(-h/38.0)$. Note that ADAM assumes that correlation is a function of distance only. The correlation of attributes at the same location is identified by the symbol 'r0':

r0(G,H): -0.73;

Specification of only r0 is sufficient when the operation considered is a point operation. Note that there may be a discontinuity in the cross-correlation at distance zero ($h = 0$).

The GIS operation for which the propagation of errors is evaluated is defined in MDL using the symbol 'calc'. For instance, when two attributes I and J are multiplied to yield the attribute K, then in MDL this is declared as:

calc("k.est","k.sdv"): I * J;

The syntax of MDL contains many more options that are not included in the examples given but that are worth mentioning here. These are options to compute skewness, kurtosis and quantiles of the output distribution and an option to obtain information on the contribution of error sources. Several language constructions have been included to increase flexibility and allow explorative use of the error propagation tool. Libraries of standard operations can be included in a model, such as Slope and Aspect for DEM analysis (see section 5.2), or membership functions for continuous classification (see section 5.4). Neighbourhood operations are implemented using a two-dimensional index of translation. As yet ADAM cannot handle error propagation with global operations, but these can be approached using other software packages (Deutsch and Journel, 1992; Pebesma and Wesseling, 1998).

The compiler

After the error propagation problem has been declared in MDL it is submitted to the ADAM compiler. The main purpose of the compiler is to generate a string of GIS operations that effectuate the error propagation. However, it also examines whether the model description is grammatically correct and whether the correlation matrix of the model parameters is positive-definite. The compiler also assists in selecting an appropriate error propagation technique, using simple rules such as:

- The Monte Carlo method is exact but time consuming.
- The Monte Carlo method can only be used if the joint distribution of all stochastic parameters is given.
- The first order Taylor method is usually acceptable for linear or sufficiently smooth models.
- The second order Taylor method is suitable for second order polynomials (ADAM can only apply this method when the inputs are jointly normal, because it uses (4.11) to compute the output variance).
- Rosenblueth's method is an alternative to the Taylor method when the model is not continuously differentiable.
- The Monte Carlo method must be used when skewness, percentiles or quantiles are required.

The selection of the optimal method can thus partly be automated (some methods do not apply to all operations) and partly depends on the user's preferences (trade-off between accuracy and computing costs).

After an appropriate error propagation technique has been chosen the compiler generates a string of GIS operations. This string of operations must be executed by the GIS to yield the actual output of the error analysis.

To derive the error propagation string for the Taylor methods requires the first and second order derivatives of the operation. For this ADAM employs a symbolic differentiation approach (Rall, 1981). There are two important advantages of symbolic differentiation over numerical differentiation. Firstly, it avoids the numerical instabilities of numerical differentiation. Secondly, it is much less demanding computationally. To appreciate fully the second advantage it is important to realise that symbolic differentiation needs to be done only once, whereas numerical differentiation needs to be repeated for every grid cell or polygon in the map.

The string of GIS operations that is generated can be further optimised by algebraic simplification, common sub-expression elimination, reduction in strength and by minimising the number of spatial operations (see Wesseling and Heuvelink, 1991).

7.3 Running the Allier case study with ADAM

The Allier example of section 5.3 is used to demonstrate how ADAM is applied in practice. The regression model is declared in MDL as follows:

```
b0: random(normal, −0.263,0.031);
b1: random(normal,0.408,0.096);
b2: random(normal,0.491,0.078);

r0(b0,b1): −0.221;
r0(b0,b2): −0.587;
r0(b1,b2): −0.655;

moistrfc: random(normal,"moistrfc.est","moistrfc.sdv");
porosity: random(normal,"porosity.est","porosity.sdv");

r0(moistrfc,porosity): "cor_fcpo.map";
r0(moistrfc,$): 0.0;
r0(porosity,$): 0.0;

calc("moistrwp.est","moistrwp.sdv"): b0 + b1*moistrfc + b2*porosity;
```

In the first three lines the regression coefficients b0, b1 and b2 are defined. These are assumed normally distributed with means and standard deviations as given. Next their correlations are given. The input maps containing the block cokriging predictions and prediction error standard deviations of porosity and moisture content at field capacity are identified by the symbolic names "moistrfc.est", "moistrfc.sdv", "porosity.est" and "porosity.sdv". The symbol '$' is a wildcard that stands here for all three regression coefficients. Note that the only type of correlation needed is the correlation at the same location. The map of correlations of the cokriging errors, identified here by the name "cor_fcpo.map", can be calculated using gstat (Pebesma and Wesseling, 1998). The last line defines the regression model. It also states that the mean and standard deviation of the model output should be stored in the maps "moistrwp.est" and "moistrwp.sdv".

When ADAM is called it advises the user to select the second order Taylor method. This is because the model is a quadratic function, all inputs are normally distributed and the only output requested is the mean and standard deviation of the moisture content at wilting point. When ADAM is run using the second order Taylor method, its output is as follows:

```
moistrwp.est = cont((−0.263 + (0.408*moistrfc.est)) + (0.491*porosity.est));
tmp0.$$$ = moistrfc.est*0.096;
tmp1.$$$ = porosity.est*0.078;
tmp2.$$$ = 0.408*moistrfc.sdv;
tmp3.$$$ = 0.491*porosity.sdv;
tmp4.$$$ = (((((0.031*tmp0.$$$)* − 0.221) + ((0.031*tmp1.$$$)* − 0.587))
    + ((tmp0.$$$*tmp1.$$$)* − 0.655))
    + ((tmp2.$$$*tmp3.$$$)*cor_fcpo.map);
tmp5.$$$ = sqr(0.096)*sqr(moistrfc.sdv);
tmp6.$$$ = ((0.096*0.078)*
    − 0.655)*((moistrfc.sdv*porosity.sdv)*cor_fcpo.map);
tmp7.$$$ = sqr(0.078)*sqr(porosity.sdv);
moistrwp.sdv = cont(sqrt((((tmp7.$$$ + (tmp6.$$$ + (tmp7.$$$ + (tmp6.$$$
    + (tmp6.$$$ + (tmp5.$$$ + (tmp6.$$$ + (tmp5.$$$ + (tmp7.$$$
    + (tmp6.$$$
    + ((tmp6.$$$ + (tmp6.$$$ + (tmp5.$$$ + (tmp5.$$$ + tmp6.$$$)))))
    + tmp7.$$$))))))))))*0.25) + ((2.0*tmp4.$$$)
    + ((((sqr(0.031) + sqr(tmp0.$$$)) + sqr(tmp1.$$$)) + sqr(tmp2.$$$))
    + sqr(tmp3.$$$))))));
```

These are the operations that must be executed by the GIS to calculate the output, i.e. the layers "moistrwp.est" and "moistrwp.sdv". The number of GIS operations is surprisingly modest considering the complexity of the second order Taylor method. This is mainly due to the combination of symbolic differentiation and algebraic simplification, because many terms evaluate to zero due to zero correlation and zero second order derivatives. Common subexpression elimination causes the creation of some temporary layers, denoted here by "tmp0.$$$" through to "tmp7.$$$".

When the first order Taylor method is used, the output of ADAM simplifies to:

```
moistrwp.est = cont((−0.263 + (0.408*moistrfc.est)) + (0.491*porosity.est));
```

tmp0.\$\$\$ = moistrfc.est*0.096;
tmp1.\$\$\$ = porosity.est*0.078;
tmp2.\$\$\$ = 0.408*moistrfc.sdv;
tmp3.\$\$\$ = 0.491*porosity.sdv;
tmp4.\$\$\$ = (((((0.031*tmp0.\$\$\$)* − 0.221) + ((0.031*tmp1.\$\$\$)* − 0.587))
  + ((tmp0.\$\$\$*tmp1.\$\$\$)* − 0.655))
  + ((tmp2.\$\$\$*tmp3.\$\$\$)*cor_fcpo.map);
moistrwp.sdv = cont(sqrt((((2.0*tmp4.\$\$\$) + ((((sqr(0.031) + sqr(tmp0.\$\$\$))
  + sqr(tmp1.\$\$\$)) + sqr(tmp2.\$\$\$)) + sqr(tmp3.\$\$\$)))));

Computational demands are somewhat less in this case. However, recall from chapter 4 that the first order Taylor method involves approximation errors. This is also the case with Rosenblueth's method, which in this case requires the evaluation of more than 90 operations involving 64 temporary layers. The string of operations is therefore not given here. The Allier example thus confirms that Rosenblueth's method is not very suitable when the number of inputs is large (here $m = 5$).

Computation of the output maps

The string of GIS operations that are the output of ADAM are in a syntax understandable by the PCRaster package (Van Deursen, 1995; Van Deursen and Wesseling, 1995; Wesseling *et al.*, 1996). PCRaster is a raster GIS developed for personal computers that is particularly useful for environmental modelling because it has added to the standard GIS functionality possibilities for dynamic modelling, geostatistical modelling and error propagation (the latter through the use of ADAM). Thus ADAM is in effect embedded in PCRaster and uses PCRaster as its processor.

However, it is an easy task to transform the output of ADAM, i.e. the string of GIS operations, to a format such that it is understandable by other GISs as well. Most vector GISs store attribute information in tables of a Relational Data Base Management System. These attribute tables can be manipulated with SQL-like languages. In order for ADAM to work with this type of GIS it should return a string of SQL-statements that, upon execution, create and compute a new attribute table containing fields of "moistrwp.est" and "moistrwp.sdv". For an example of the use of ADAM in combination with vector GIS, see Wesseling and Heuvelink (1993b).

Note that the principle of separating the compiler (ADAM) from the processor (the GIS) proves its merit here. This approach gives considerable flexibility, simply because ADAM does not presuppose a specific GIS data structure. Consequently, there are no serious obstacles for linking ADAM with any of the many GISs that are nowadays on the market.

Other software packages that have error propagation functionality are UNCSAM (Janssen *et al.*, 1994), UNCERT (Wingle *et al.*, 1996) and IDRISI (Eastman, 1997).

CHAPTER EIGHT

Summary and conclusions

8.1 Summary of research results: the list of nine research questions

This monograph began with a list of nine research questions concerning the propagation of quantitative attribute errors in spatial and environmental modelling with GIS. These questions have all been addressed in the foregoing chapters. Let me begin this chapter by reviewing them one by one:

1 What is a suitable definition of error for quantitative spatial attributes?

A stochastic model was formulated to represent the error in quantitative spatial attributes. The rationale behind the choice for a stochastic model was that because of uncertainty, the best one can do is to represent the value of a spatial attribute at any given location by a distribution of possible values, and not by a single deterministic value.

The stochastic model of error contains many parameters, such as the mean – representing systematic error or bias, and the standard deviation – representing non-systematic, random error. These parameters are functions defined on a geographical domain. The spatial dimension of the problem also requires that a full characterisation of the error model must include spatial autocorrelation. When multiple attributes are considered, (spatial) cross-correlation must be defined as well. Once the distribution type of the error has also been specified, the stochastic error model satisfies the criterion set by Goodchild *et al.* (1992), namely that an error model should be 'a stochastic process capable of generating a population of distorted versions of the same reality'.

When a stochastic model is adopted to represent the error in a quantitative spatial attribute then this means that the error is represented as a random field. Random fields are the central object of study in the field of spatial statistics. Consequently, an important advantage of characterising spatial errors by random fields is that use can be made of the vast and well-developed theory on random fields, as contained in such reference books as Journel and Huij-bregts (1978) and Cressie (1991). Moreover, new developments in this field, such as reported in scientific journals and in conference proceedings (e.g. Soares, 1993; Baafi and Schofield, 1997), can be used to improve or extend the error model.

2 Given this definition of error, how should the error for a particular input attribute to a GIS operation be identified?

Certain stationarity assumptions are needed to estimate the parameters of the error model in practice. In many situations it is more realistic to impose these assumptions on the spatial variability of the attribute, rather than on its uncertainty. The uncertainty or error in a spatial attribute depends on the spatial variability, but it is also influenced by the mapping procedure used.

The spatial variability of a spatial attribute was characterised by the general model of spatial variation (GMSV). By making further assumptions on the GMSV it can be simplified to the discrete model of spatial variation (DMSV), the continuous model of spatial variation (CMSV) or the mixed model of spatial variation (MMSV).

By including the mapping procedure in the subsequent identification procedure the parameters of the error model can be identified from the chosen model of spatial variation. In the case of the CMSV this is done by kriging or, in the multivariate situation, by cokriging. An important problem that turns up here is that the resulting spatial auto- and cross-correlations of the error random field are usually not a function of only the distance between locations. This may cause considerable storage problems in practice. One way around this problem is to use conditional simulation in combination with the Monte Carlo method (see below).

The identification of the error model may also be based on expert judgement or heuristic guesses. This will be the case when the mapping procedure cannot easily be translated into mathematical terms, such as is the case in a free soil survey (Bregt, 1992), and when quantitative measures of within-map unit variability are lacking. However, empirical evidence shows that relying on informed guesses is not without risk and should be treated with care (Heuvelink and Bierkens, 1992).

3 Which techniques can be used to analyse the propagation of errors through local and global GIS operations?

Analytical results for the error propagation problem can only be obtained in a few special cases, such as when the GIS operation is a linear function. Numerical solutions are not very attractive either, and so four alternative approaches are suggested instead.

The first order Taylor method approaches the problem by approximating the GIS operation by a linear function. The linearisation is done in a way such that approximation errors are relatively small for input values that have a larger probability of occurrence. The linearisation makes the problem analytically tractable, so that the method yields analytical results that can provide useful insight into the error propagation process.

The second order Taylor method is a logical extension of the first order Taylor method. It uses a quadratic function to approximate the operation. This presumably leads to smaller approximation errors, at the expense of an increase in complexity.

Rosenblueth's method is an alternative to the first order Taylor method when the GIS operation is not continuously differentiable. It approximates the operation using a 'smoothed' derivative rather than the usual 'point' derivative, as used by the first order Taylor method. The resulting equations are not as transparent as the corresponding Taylor equations.

The Monte Carlo method uses a simulation approach to analyse the propagation of errors in GIS operations. It repeatedly draws realisations from the joint probability distribution of the input attributes, each time substituting these realisations into the operation, computing the result and storing it. In this way a random sample from the output distribution is obtained, which is analysed using techniques from classical sampling theory.

All four methods of error propagation just described have their drawbacks (see below). Some of the methods do not apply to all operations, others can become extremely time-consuming or may involve large approximation errors. However, in many cases there will be at least one method that is appropriate for a given situation. Thus the methods are in a sense complementary, and as a group in almost all cases they enable one successfully to trace the propagation of errors in quantitative spatial modelling with GIS.

4 What are the advantages and disadvantages of the various error propagation techniques?

The main problem with the Taylor methods is that the results are approximate only. This is also true for Rosenblueth's method. It will not always be easy to determine whether the approximations involved using these techniques are acceptable. The Monte Carlo method does not suffer from this problem because it can reach an arbitrary level of accuracy.

The Monte Carlo method brings along other problems, though. High accuracies are reached only when the number of runs is sufficiently large, which may cause the method to become extremely time-consuming. Another disadvantage of the Monte Carlo method is that the results do not come in an analytical form.

As a general rule it seems that the Taylor methods may be used to obtain crude preliminary answers. These should provide sufficient detail to be able to obtain at least an indication of the intrinsic quality of the output of the GIS operation. In fact, this will be sufficient for most purposes, given that the order of magnitude of errors is often of more interest than exact values. When exact values or quantiles and/or percentiles are needed the Monte Carlo method may be used.

The Monte Carlo method will probably also be preferred when error propagation with complex operations is studied. This is because the method is easily implemented and generally applicable. In particular, it seems that for global operations the Monte Carlo method is most suitable. This then requires the stochastic simulation of random fields, for which a number of techniques can be applied. The use of conditional simulation from the stationary random field $Z(\cdot)$ offers a possible solution to the problem with storing the error parameters of the non-stationary error random field $V(\cdot)$.

5 How do these error propagation techniques perform when they are applied to practical problems?

The theory was demonstrated using four case studies. In each of these case studies one or more error propagation techniques were applied, and a comparison of the performance of the techniques was made in the first case study. This may not be sufficient to reach conclusive results, but some useful conclusions can be drawn here.

The case studies concern relevant practical problems. In all of these cases it was relatively easy to apply the error analysis using one of the four error propagation techniques. In other words, the case studies suggest that there are no fundamental obstacles to employing error propagation techniques within GIS on a routine basis.

In some cases the Monte Carlo method required a large number of runs to reach stable results. This was particularly true for the Geul case study, where one of the inputs was lognormally distributed. The computing time used by the other three methods was much less. From this it can be concluded that computing limitations should not be a serious obstacle to the application of error propagation techniques to standard GIS operations.

The case studies also showed that there is much more to be gained from an error analysis than only the computation of output error. For instance, the Geul case study also revealed that it is imperative to pay attention to support issues, when the objective is to estimate the probability that the Acceptable Daily Intake is exceeded. The Balazuc case study showed the effect of an increase in spatial autocorrelation on the errors in derived products from a digital elevation model. For the Allier case study it was found that model error was small in comparison with input error, so that rational decisions could be made as to how best to reduce output error. The Lacombe case study was used to compare the robustness of Boolean and continuous classification with errors in input attributes. These are just a few examples of how an error analysis can be used to glean valuable information about how uncertainty aspects affect the output of computational models.

This monograph has not reported a detailed example of how error propagation can be applied to global operations, in particular to distributed dynamic models. Such an analysis requires a thorough knowledge of the model used and can best be done in close collaboration with experts from the discipline to which the model applies. Some examples of the use of the Monte Carlo method with models involving global operations and/or dynamic distributed models are Beven and Binley (1992), De Roo *et al.* (1992), Rossi *et al.* (1993), Gotway (1994) and Binley *et al.* (1997).

6   When the GIS operation is in effect a computational model, how should model error be incorporated in the error propagation analysis?

Model error can easily be included in the error analysis by extending the number of stochastic inputs to the GIS operation. An example is given by the Allier case study, in which model error was included by defining the regression coefficients as random variables. The error analysis itself does not distinguish between input and model error; both are represented as random variables or random fields.

Usually, the main problem with including model error in the analysis is its identification, especially when it concerns complex conceptual models. Even when model error is represented by a lumped residual noise term, its quantification will not always be easy.

7   When multiple error sources cause error in the output of a GIS operation, how can the relative contributions of the error sources be determined?

The contribution of individual error sources can be obtained by utilising the partitioning property discussed in section 4.4. This property allows one to determine how much of the output variance can be attributed to the individual inputs, or, in the case of correlated inputs, to groups of inputs. It is also possible to determine the relative contributions of input and model error.

With the Taylor methods the relative contributions of the error sources can be obtained directly from the analytical expressions for the output variance. The contribution of individual error sources can also be obtained using the Monte Carlo method, but this requires a substantial increase in the number of Monte Carlo runs.

8   If the error in the output exceeds a critical level, what action can best be undertaken to improve accuracy? Conversely, if the quality of the output exceeds requirements by far, how can savings in data collection and processing best be made?

This question can be addressed using the partitioning property just discussed. Clearly the largest reduction in output error can be achieved by an improved mapping of the input that has the largest error contribution. However, the cost of mapping should also be included in deciding how to act. Some attributes are more expensive to measure than others, so that a cost–benefit analysis may show that for a given budget more can be gained from the improved mapping of another input attribute. Knowledge of the variogram can be used to optimise sampling (McBratney *et al.*, 1981; Burrough, 1991).

Note also that it may sometimes prove practically impossible to reduce the error of a particular input attribute. For instance, no matter how many observations are collected in the field, the point kriging variance at an unvisited location will always be at least as large as the nugget, unless perhaps a different support or measurement technique is used.

When model error is the major source of error, then ways must be sought to improve the model. This will often be difficult and costly. It can be done by improving the calibration or by using different, more complicated models (Burrough, 1992b; Heuvelink, 1998).

The error analysis can also show which input errors hardly affect the quality of the output, so that savings on mapping can be made with negligible consequences. As a general rule, it is thus best to strive for a balance of errors.

9   How should error propagation techniques be implemented in a GIS?

Several points must be addressed when the implementation of error propagation techniques is considered. First, the error propagation tool should not reinvent the GIS. Instead, it should make use of the existing database structure and analytical functionality of the GIS. Secondly, the tool should be sufficiently flexible so that it can easily be linked to various types of GISs, be it raster or vector. Thirdly, the user interface deserves special attention because an error analysis requires many statistical parameters, which the user must supply. Fourthly, attention must also be paid to performance issues, because a clever implementation can greatly improve the efficiency of the numerical analysis.

We developed a software tool, named ADAM, that takes all four points into account. An important and integral part of ADAM is the model description language (MDL), in which the user defines the model, its inputs and their statistical parameters. When ADAM is run it assists the user to select the most appropriate error propagation technique. ADAM does not perform any calculations by itself, it merely translates the problem into a sequence of standard GIS operations, that must be executed by the GIS. Separating the compiler (ADAM) from the processor (the GIS) was done on purpose to increase flexibility. At present ADAM is included in the PCRaster package (Van Deursen and Wesseling, 1995), but a link with other GISs can easily be made.

It is important that concrete steps are taken to incorporate error propagation techniques, such as contained in ADAM, in commercial GIS. Only then will these techniques become available to a larger public. Moreover, the inte-

gration of error propagation techniques and GIS will also increase the aware-
ness of users to the problem of errors and error propagation in GIS. Ideally,
each GIS should standardly contain a general purpose 'error button'
(Openshaw et al., 1991).

8.2   Summary of research results: additional results

Although not on the list of research questions, this research yielded several
other results that are worth reviewing here.

The role of the support size

One thing that became clear in chapter 2 is that when spatial attributes are
accompanied by their uncertainties, then it is important that the support of the
attributes is reported as well. This is because the uncertainty is strongly
support-dependent. For instance, the uncertainty of block-averaged values is
usually less than that of point values, due to an averaging-out effect. Consequently, when the propagation of errors in a GIS operation with multiple
inputs is considered, care must be taken that the supports of the inputs
mutually agree.

When a change of support is carried out, then the results are sensitive to the
assumptions of the adopted model of spatial variation. In some cases this may
yield unrealistic results, so that it were better if a change of support be
avoided. Unfortunately, this is not always possible. The case studies contain
several examples that demonstrate this particular problem.

When the support is that large that a feasible option is to estimate attribute
errors using a design-based approach, such as is the case when map-unit averages are required, then this approach should in many cases be preferred over
the model-based approach. Whether a model-based or design-based approach
is used does not directly affect the error propagation process itself, but it has a
large impact on the identification of input error.

The mixed model of spatial variation

In chapters 2 and 3 the mixed model of spatial variation (MMSV) was introduced and applied to a case study. The mixed model of spatial variation is
more attractive than its discrete and continuous counterparts because it can
handle both abrupt and gradual spatial variability present in the same area. In
addition, it turns out that the MMSV performs well on the whole range of
spatial variation, ranging from purely discrete to purely continuous
(Heuvelink and Huisman, 1996). This is important because it frees the user

from an ambiguous choice beforehand between abrupt and gradual spatial variation.

Assessing the accuracy of regression models for mapping expensive-to-measure spatial attributes

The Allier case study showed an application that appears to have general validity. In general terms, the procedure can be described as follows:

1 The objective is to map a spatial attribute that is expensive-to-measure.

2 The expensive attribute is correlated with cheaper-to-measure attributes.

3 The relationship between the expensive and cheap attributes can be quantified by a regression model, coefficients of which are estimated from a limited number of observations of all attributes. Due to budget constraints the number of observations is not very large.

4 The regression model predicts the value of the expensive attribute from given values of the cheap attributes. However, the prediction will be in error due to the presence of model error. The model error is contained in the standard deviations of the regression coefficients and in the regression residual.

5 The cheap attributes are mapped by interpolating many observations. Many observations can be collected because these measurements are cheap. Nonetheless interpolation errors occur, and these can be quantified using kriging.

6 If the regression model is accepted as valid, then a map of the expensive attribute can be obtained upon substitution of the interpolated maps into the regression equation. The result will be in error due to model error as well as interpolation error.

7 Error propagation analysis can be used to determine how the error sources combine to yield the error in the map of the expensive attribute, and to determine which error source has the largest contribution.

The procedure is restated here because it seems applicable to many practical situations. Pedotransfer functions are often used (Cosby *et al.*, 1984; Bouma, 1989; Burrough, 1989b; Vereecken *et al.*, 1989; Van Diepen *et al.*, 1991; Finke *et al.*, 1998), and more attention is now being paid to the uncertainty contained in these functions (Kros *et al.*, 1992; Finke *et al.*, 1996).

It is worth mentioning that the procedure outlined above may be improved by adapting the regression to the fact that the input maps are more erroneous than the data used to calibrate the regression (Elston *et al.*, 1997). Another interesting research question is to study how the procedure described above compares to alternative methods, such as cokriging (Stein *et al.*, 1988; Cressie, 1991). Cokriging will probably be favoured when there are sufficient observations to obtain the variogram (and cross-variograms) of the expensive attribute. Note that the regression procedure outlined above implicitly imposes a variogram on the expensive attribute. It can easily be derived from the variograms of the cheap attributes and the regression model.

Continuous classification is more robust to error propagation than Boolean classification

When the inputs to a logical model are uncertain, the output of the model will be in error as well. However, the propagation of errors is not the same for Boolean and continuous classification. Section 5.4 clearly shows that Boolean methods of logical modelling are much more prone to error propagation than the continuous equivalents.

Continuous classification is thus more robust to errors in the input attributes. Also, many environmental problems cannot realistically be modelled by the rigid Boolean classification rules. These two advantages of continuous classification should cause users of Boolean classification systems used for environmental data to re-examine the basis of their ideas in order to see how the minor modification of introducing continuous classes with a transition zone would help them to make better use of their data.

The iterative method for autoregressive spatial simulation

In chapter 6 an iterative method was presented to simulate autoregressive random fields. The method is straightforward, generally applicable and it avoids the computational overhead associated with the traditional inversion method. The convergence rate of the method is sensitive to the degree of spatial autocorrelation of the spatial autoregressive model. Numerical experiments show that the computational load of the method is modest and allows the simulation of grids of respectable size.

8.3   Towards a full grown error handling capability of GIS

There remain several problems concerning the propagation of errors in quantitative spatial modelling that deserve additional attention. Some of these have merely been touched upon in the course of this research, others have as yet been completely ignored.

Theoretical research on error propagation

First of all it is useful to examine more closely the approximation error induced by the Taylor method. If general statements can be made on this particular problem, such as the specification of maximum error bounds, then this would allow one to decide in advance whether the Taylor method is acceptable for a given situation. This problem may be solved by analysing the properties of the remainder of the Taylor series (Casella and Berger, 1990), or possibly using Beale's nonlinearity measure (Kuczera, 1988). Another problem is to decide how many Monte Carlo runs are needed to obtain a given level of accuracy. Some attention has been given to this problem in section 4.2.4, but this particular point deserves more attention. In line with this problem a viable option seems to be to replace point estimates by interval estimates (Casella and Berger, 1990). Interval estimation yields wider intervals when the number of Monte Carlo runs is small, and so it implicitly incorporates the confidence gain resulting from an increase of runs.

In section 4.2.4 it was also shown that the efficiency of the Monte Carlo method is proportional to the square root of the number of runs. This implies that it is extremely costly to obtain results with a high level of accuracy. It seems worthwhile to examine how the 'crude' Monte Carlo method can be improved by employing various variance reduction techniques (Hammersley and Handscomb, 1979; Lewis and Orav, 1989). One possibility would be to use the result of the Taylor method as a control variate, after which the remainder may be evaluated using the crude Monte Carlo method. Another option is to employ importance or stratified sampling, such as Latin Hypercube sampling (Stein, 1987).

Another problem ignored in this research concerns the autocorrelation of the output of the GIS operation, and its cross-correlation with other attributes stored in the GIS database. Both can in principle be derived from a Monte Carlo analysis, provided a global approach with simulated random fields is used, but there remain several important practical problems to be resolved.

Identification of input and model error

An error propagation analysis starts with the premise that the parameters of the error model are correctly specified. However, in many practical situations the user has no definite information about the errors associated with the attributes that are stored in a GIS, nor does he know exactly how accurate his computational models are. This is an important problem because an error propagation analysis will only yield sensible results if the input and model errors have realistic values.

The problem of input error assessment is particularly difficult when one is confined to quantifying the error through informed guesses. Relying only on expert information can be quite risky. This is partly due to the lack of experience and training of map makers. The present situation can be improved if map makers such as soil surveyors realise that it is their duty to convey the accuracy of the maps they produce, even when accuracy is less than expected.

A more fundamental problem is to decide which model of spatial variation to use for the representation of spatial data. In chapter 3 we have seen that the decision will have a marked effect on the error identification. It is inappropriate to let the decision between the discrete, mixed and continuous models of spatial variation depend on only the type of information available. We must make sure that the right model is used with the right data, i.e. the choice between the models should be dictated by the type of spatial variability present. Note also that in some cases a more elaborate version of the GMSV may be more appropriate.

Another problem ignored in this monograph is temporal variation. Evidently, many attributes vary in time as well as in space, and so an error model for spatio-temporal data should also include temporal variability (Bennett, 1979; Cressie, 1991; Heuvelink *et al.*, 1997).

Applications of the theory

The purpose of this research was to develop and implement a methodology for tracing error propagation in GIS. Several applications were included to demonstrate the theory, but application was not the main objective here. Now that a tool is available it should be applied to relevant practical problems. Many more case studies are needed to fully appreciate the value of error propagation techniques for spatial modelling with GIS.

Although as yet the use of error propagation in GIS is far from a routine exercise, within the environmental sciences uncertainty analyses are by now quite common. Recent examples from various branches within the environmental sciences are Kros *et al.* (1992), Rossi *et al.* (1993), Bolstad and Stowe (1994), De Jong (1994, chapter 9), Gotway (1994), Jansen *et al.* (1994), Leenhardt (1995), Finke *et al.* (1996), Woldt *et al.* (1996), Binley *et al.* (1997), Dobermann and Oberthür (1997), Hunter and Goodchild (1997) and Van Der Perk (1997).

The development of a GIS for uncertain spatial data

The aim of this research was to analyse the propagation of errors in quantitative spatial modelling with GIS. In so doing, the GIS was treated as an externally defined system that is used to store the data, to carry out the calculations and to display the results. This is also the approach followed by the ADAM error propagation tool, because it operates independently from the GIS. The purpose was to extend the GIS with error propagation functionality, not to change the GIS itself.

At present the GIS is thus 'unaware' of the fact that it operates with uncertain spatial data. This approach works satisfactorily in practice but it has its limitations. A better solution would be if the GIS were to deal directly and explicitly with uncertain spatial attributes. This would be a GIS for uncertain data, attributes of which are stochastic rather than deterministic. Such a GIS should adapt its database structure to the specific properties of stochastic attributes. For instance, most current GISs assume that a spatial attribute is identified by a single data layer or a single column in a relational data base management system. But stochastic attributes are usually identified by at least two layers or columns, such as the mean and standard deviation (Wesseling and Heuvelink, 1993b).

Stochastic attributes also have several other characteristic parameters, such as auto- and cross-correlation. These must also be represented in the GIS. To avoid storage problems of correlations for non-stationary attributes, as was mentioned in chapter 2, it may be more sensible to store the original point observations, rather than the maps that have been derived from them. Maps of predictions, prediction error standard deviations and correlations can then be made upon request. Clearly, such a GIS should also contain the statistical tools that are needed to perform geostatistical interpolation and conditional simulation. Such a GIS must also find a satisfactory way of handling the large amounts of data that result from a Monte Carlo analysis. Finally, it must also find efficient ways to visualise the uncertainty about spatial data (Beard *et al.*, 1991; Hearnshaw and Unwin, 1994; Klinkenberg and Joy, 1994).

It will be clear that current GISs must undergo some fundamental changes if they are to handle and manage uncertain spatial data professionally. It is also clear that the poor functionality of current GISs is an important obstacle against the large-scale introduction of error propagation analyses in GIS. However, equally important obstacles are the lack of information about the accuracy of source data stored in and computational models linked to the GIS and the fact that the majority of GIS users as yet do not recognise the added value of an error propagation analysis. Performing an uncertainty analysis asks a lot of the user. The user must know something about the theory of error propagation and carrying out an error analysis is often quite laborious. Therefore users will not carry out an error analysis unless they see themselves that the analysis serves a clear goal (Fisher, 1997). With this monograph I hope to have contributed to making users more aware of the usefulness of error propagation analyses in GIS.

References

ABBOTT, M. B., BATHURST, J. C., CUNGE, J. A., O'CONNEL, P. E. and RASMUSSEN, J., 1986, An introduction to the European Hydrological System – Système Hydrologique Européen: SHE, 2. Structure of a physically-based, distributed modeling system, *Journal of Hydrology*, **87**, 61–77.

AHO, A. V., SETHI, R. and ULLMAN, J. D., 1986, *Compilers: Principles, Techniques and Tools*, Reading: Addison-Wesley.

ALABERT, F. 1987, The practice of fast conditional simulations through the LU decomposition of the covariance matrix, *Mathematical Geology*, **19**, 369–86.

ANSELIN, L., DODSON, R. F. and HUDAK, S., 1994, Linking GIS and spatial data analysis in practice, *Geographical Systems*, **1**, 3–23.

ARMSTRONG, M., 1984, Problems with universal kriging, *Mathematical Geology*, **16**, 101–8.

BAAFI, E. Y. and SCHOFIELD, N. A. (Eds.), 1997, *Geostatistics Wollongong '96*, Dordrecht: Kluwer.

BAKER, R., 1984, Modeling soil variability as a random field, *Mathematical Geology*, **16**, 435–48.

BEARD, M. K., BUTTENFIELD, B. P. and CLAPHAM, S. B., 1991, Visualization of spatial data quality, Technical paper 91-26, National Center for Geographic Information and Analysis, University of Maine.

BEASLEY, D. B. and HUGGINS, L. F., 1980, ANSWERS: A model for watershed planning, *Transactions of the ASAE*, **23**, 938–44.

BECKETT, P. H. T. and BURROUGH, P. A., 1971, The relation between cost and utility in soil survey. IV: Comparison of the utilities of soil maps produced by different survey procedures, and to different scales, *Journal of Soil Science*, **22**, 466–80.

BELLIN, A., SALANDIN, P. and RINALDO, A., 1992, Simulation of dispersion in heterogeneous porous formations: statistics, first-order theories, convergence of computations, *Water Resources Research*, **28**, 2211–27.

BENNETT, R. J., 1979, *Spatial Time Series, Analysis-Forecasting-Control*, London: Pion.

BEVEN, K., 1989, Changing ideas in hydrology – the case of physically-based models, *Journal of Hydrology*, **105**, 157–72.

BEVEN, K. and BINLEY, A., 1992, The future of distributed models: model calibration and uncertainty prediction, *Hydrological Processes*, **6**, 279–98.

BIERKENS, M. F. P. and BURROUGH, P. A., 1993a, The indicator approach to categorical soil data. I. Theory, *Journal of Soil Science*, **44**, 361–8.

BIERKENS, M. F. P. and BURROUGH, P. A., 1993b, Modelling of map impurities using sequential indicator simulation, in SOARES (1993), pp 637–48.

BINLEY, A., BUCKLEY, K., CALORE, C., ULDERICA, P. and LA BARBERA, P., 1997, Modelling uncertainty in estimates of recharge to a shallow coastal aquifer, *Hydrological Sciences*, **42**, 155–68.

BOLSTAD, P. V. and STOWE, T., 1994, An evaluation of DEM accuracy: elevation, slope and aspect, *Photogrammetric Engineering & Remote Sensing*, **60**, 1327–32.

BOLSTAD, P. V., GESSLER, P. and LILLESAND, T. M., 1990, Positional uncertainty in manually digitized map data, *International Journal of GIS*, **4**, 399–412.

BORGMAN, L., TAHERI, M. and HAGAN, R., 1984, Three-dimensional, frequency-domain simulations of geological variables, in VERLY, G., DAVID, M., JOURNEL, A. G. and MARECHAL, A. (Eds.) *Geostatistics for Natural Resource Characterization*, Vol. 1, pp. 517–41, Dordrecht: Reidel.

BOUMA, J., 1989, Using soil survey data for quantitative land evaluation, in STEWART, B. A. (Ed.) *Advances in Soil Science*, **9**, 177–213.

BOUTEN, W., 1993, Monitoring and modelling forest hydrological processes in support of acidification research, PhD thesis, University of Amsterdam, The Netherlands.

BREGT, A. K., 1992, Processing of soil survey data, PhD thesis, Wageningen Agricultural University, The Netherlands.

BREGT, A. K. and BEEMSTER, J. G. R., 1989, Accuracy in predicting moisture deficits and changes in yield from soil maps, *Geoderma*, **43**, 301–10.

BROOKER, P. I., 1985, Two-dimensional simulation by turning bands, *Mathematical Geology*, **17**, 81–90.

BURROUGH, P. A., 1986, *Principles of Geographical Information Systems for Land Resources Assessment*, Oxford: Clarendon Press.

BURROUGH, P. A., 1989a, Fuzzy mathematical methods for soil survey and land evaluation, *Journal of Soil Science*, **40**, 477–92.

BURROUGH, P. A., 1989b, Matching spatial databases and quantitative models in land resource assessment, *Soil Use and Management*, **5**, 3–8.

BURROUGH, P. A., 1991, Sampling designs for quantifying map unit composition, in MAUSBACH, M. and WILDING, L. (Eds.) *Spatial Variabilities of Soils and Landforms*, pp. 89–125, Madison: Soil Science Society of America Special Publication no. 28.

BURROUGH, P. A., 1992a, Are GIS data structures too simple minded?, *Computers and Geosciences*, **18**, 395–400.

BURROUGH, P. A., 1992b, The development of intelligent geographical information systems, *International Journal of GIS*, **6**, 1–11.

BURROUGH, P. A., 1993, Soil variability: a late 20th century view, *Soils and Fertilizers*, **56**, 529–62.

BURROUGH, P. A., MACMILLAN, R. A. and VAN DEURSEN, W., 1992, Fuzzy classification methods for determining land suitability from soil profile observations and topography, *Journal of Soil Science*, **43**, 193–210.

BURROUGH, P. A. and MCDONNELL, R. A., 1998, *Principles of Geographical Infor-*

mation Systems, Oxford: Oxford University Press.

CARTER, J. R., 1992, The effect of data precision on the calculation of slope and aspect using gridded DEMs, *Cartographica*, **29**, 22–34.

CASELLA, G. and BERGER, R. L., 1990, *Statistical Inference*, Pacific Grove: Wadsworth and Brooks/Cole.

CASPARY, W. and SCHEURING, R., 1992, Error-bands as measures of geometrical accuracy, in HARTS, J., OTTENS, H. F. L. and SCHOLTEN, H. J. (Eds.) *Proceedings EGIS '92*, pp. 226–33, Utrecht: EGIS Foundation.

CHRISMAN, N. R., 1989, Modeling error in overlaid categorical maps, in GOODCHILD and GOPAL (1989), pp. 21–34.

CHRISTAKOS, G., 1987, Stochastic simulation of spatially correlated geo-processes, *Mathematical Geology*, **19**, 807–31.

CHRISTENSEN, R., 1991, *Linear Models for Multivariate, Time Series, and Spatial Data*, New York: Springer.

CINLAR, E., 1975, *Introduction to Stochastic Processes*, Englewood Cliffs, New Jersey: Prentice-Hall.

COCHRAN, W. G., 1977, *Sampling Techniques*, 3rd Edn, New York: Wiley.

CONGALTON, R. G., 1988, A comparison of sampling schemes used in generating error matrices for assessing the accuracy of maps generated from remotely sensed data, *Photogrammetric Engineering & Remote Sensing*, **54**, 593–600.

COSBY, B. J., HORNBERGER, G. M., CLAPP, R. B. and GINN, T. R., 1984, A statistical exploration of the relationships of soil moisture characteristics to the physical properties of soils, *Water Resources Research*, **20**, 682–90.

CRESSIE, N., 1991, *Statistics for Spatial Data*, New York: Wiley.

CRESSIE, N., 1993, Aggregation in geostatistical problems, in SOARES (1993), pp. 25–36.

CURRAN, P. J., 1985, *Principles of Remote Sensing*, Harlow, United Kingdom: Longman.

DAVIS, J. C., 1986, *Statistics and Data Analysis in Geology*, New York: Wiley.

DAVIS, M. W., 1987a, Production of conditional simulations via the LU triangular decomposition of the covariance matrix, *Mathematical Geology*, **19**, 91–8.

DAVIS, M. W., 1987b, Generating large stochastic simulations – the matrix polynomial approximation method, *Mathematical Geology*, **19**, 99–107.

DCDSTF, DIGITAL CARTOGRAPHIC DATA STANDARDS TASK FORCES, 1988, The proposed standard for digital cartographic data, *The American Cartographer*, **15**, 1–140.

DE GRUIJTER, J. J. and TER BRAAK, C. J. F., 1990, Model-free estimation from spatial samples: a reappraisal of classical sampling theory, *Mathematical Geology*, **22**, 407–15.

DE JONG, S. M., 1994, Applications of reflective remote sensing for land degradation studies in a Mediterranean environment, PhD thesis, University of Utrecht, The Netherlands.

DE JONG, S. M. and RIEZEBOS, H. TH., 1988, Erosierisicokartering met behulp van GIS – een voorbeeld uit de Camargue (Zuid-Frankrijk), in BERENDSEN, H. J. A. and VAN STEIJN, H. (Eds.) *Nieuwe Karteringsmethoden in de Fysische Geografie*, pp. 45–53, Amsterdam/Utrecht: Nederlandse Geografische Studies 63 (in Dutch).

DELHOMME, J. P., 1979, Spatial variability and uncertainty in groundwater flow parameters: a geostatistical approach, *Water Resources Research*, **15**, 269–80.

DE ROO, A. P. J., 1993, Modelling surface runoff and soil erosion in catchments using Geographical Information Systems, PhD thesis, University of Utrecht, The Netherlands.

DE ROO, A. P. J., HAZELHOFF, L. and BURROUGH, P. A., 1989, Soil erosion modelling using ANSWERS and geographical information systems, *Earth Surface Processes and Landforms*, **14**, 517–32.

DE ROO, A. P. J., HAZELHOFF, L. and HEUVELINK, G. B. M., 1992, Estimating the effects of spatial variability of infiltration on the output of a distributed runoff and soil erosion model using Monte Carlo methods, *Hydrological Processes*, **6**, 127–43.

DETTINGER, M. D. and WILSON, J. L., 1981, First order analysis of uncertainty in numerical models of groundwater flow. Part I. Mathematical development, *Water Resources Research*, **17**, 149–61.

DEUTSCH, C. V. and JOURNEL, A. G., 1992, *GSLIB: Geostatistical Software Library and User's Guide*, New York: Oxford University Press.

DEVROYE, L., 1986, *Non-Uniform Random Variate Generation*, New York: Springer.

DOBERMANN, A. and OBERTHÜR, T., 1997, Fuzzy mapping of soil fertility – a case study on irrigated riceland in the Philippines, *Geoderma*, **77**, 317–39.

DOMBURG, P., DE GRUIJTER, J. J. and BRUS, D. J., 1993, A structured approach to designing soil survey schemes with prediction of sampling error from variograms, *Geoderma*, **62**, 151–64.

DOWD, P. A., 1992, A review of recent developments in geostatistics, *Computers and Geosciences*, **17**, 1481–500.

DUMANSKI, J. and ONOFREI, C., 1989, Crop yield models for agricultural land evaluation, *Soil Use and Management*, **5**, 9–16.

DUNN, R., HARRISON, A. R. and WHITE, J. C., 1990, Positional accuracy and measurement error in digital databases of land use: an empirical study, *International Journal of GIS*, **4**, 385–98.

EASTMAN, J. R., 1997, IDRISI: a grid based geographic analysis system, Worcester: Clark University (http://www.idrisi.clarku.edu/).

EDWARDS, G. and LOWELL, K. E., 1996, Modeling uncertainty in photointerpreted boundaries, *Photogrammetric Engineering & Remote Sensing*, **62**, 337–91.

ELSTON, D. A., JAYASINGHE, G., BUCKLAND, S. T., MACMILLAN, D. C. and ASPINALL, R. J., 1997, Adapting regression equations to minimize the mean squared error of predictions made using covariate data from a GIS, *International Journal of GIS*, **11**, 265–80.

FINKE, P. A., BOUMA, J. and HOOSBEEK, M. R. (Eds.), 1998, Soil and water quality at different scales, *Nutrient Cycling in Agroecosystems*, Kluwer, in press.

FINKE, P. A., WÖSTEN, J. H. M. and JANSEN, M. J. W., 1996, Effects of uncertainty in major input variables on simulated functional soil behaviour, *Hydrological Processes*, **10**, 661–9.

FISHER, P. F., 1990, Simulation of error in digital elevation models, *Papers and Proceedings of Applied Geography Conferences 13*, pp. 37–43.

FISHER, P. F., 1992, First experiments in viewshed uncertainty: simulating fuzzy viewsheds, *Photogrammetric Engineering & Remote Sensing*, **58**, 345–52.

FISHER, P. F., 1997, Book review of *Elements of Spatial Data Quality*, *International Journal of GIS*, **11**, 407–8.

FORIER, F. and CANTERS, F., 1996, A user-friendly tool for error modelling and error propagation in a GIS environment, in MOWRER, H. T., CZAPLEWSKI, R. L. and HAMRE, R. H. (Eds.) *Spatial Accuracy Assessment in Natural Resources and Environmental Sciences*, Fort Collins, Colorado: USDA Forest Service Technical Report RM-GTR-277, pp. 225-34.

GILCHRIST, W., 1984, *Statistical Modelling*, Chichester: Wiley.

GÓMEZ-HERNÁNDEZ, J. J., 1993, Regularization of hydraulic conductivities: a numerical approach, in SOARES (1993), pp. 767-78.

GÓMEZ-HERNÁNDEZ, J. J. and JOURNEL, A. G., 1993, Joint sequential simulation of multigaussian fields, in SOARES (1993), pp. 85-94.

GÓMEZ-HERNÁNDEZ, J. J. and SRIVASTAVA, R. M., 1990, ISIM3D: an ANSI-C three-dimensional multiple indicator conditional simulation program, *Computers and Geosciences*, **16**, 395-440.

GOODCHILD, M. F., 1989, Modeling error in objects and fields, in GOODCHILD and GOPAL (1989), pp. 107-13.

GOODCHILD, M. F., 1992, Geographical data modeling, *Computers and Geosciences*, **18**, 401-8.

GOODCHILD, M. F. and GOPAL, S. (Eds.), 1989, *Accuracy of Spatial Databases*, London: Taylor & Francis.

GOODCHILD, M. F., SUN, G. and YANG, S., 1992, Development and test of an error model for categorical data, *International Journal of GIS*, **6**, 87-104.

GOODCHILD, M. F., PARKS, B. O. and STAEYERT, L. T. (Eds.), 1993, *Environmental Modeling with GIS*, New York: Oxford University Press.

GOOVAERTS, P., 1992, Factorial kriging analysis: a useful tool for exploring the structure of multivariate spatial soil information, *Journal of Soil Science*, **43**, 597-619.

GOTWAY, C. A., 1994, The use of conditional simulation in nuclear-waste-site performance assessment (with discussion), *Technometrics*, **36**, 129-61.

GOULARD, M. and VOLTZ, M., 1992, Linear coregionalization model: tools for estimation and choice of cross-variogram matrix, *Mathematical Geology*, **24**, 269-86.

GRAYSON, R. B., BLÖSCHL, G., BARLING, R. D. and MOORE, I. D., 1993, Process, scale and constraints to hydrological modelling in GIS, in KOVAR and NACHTNEBEL (1993), pp. 83-92.

GRAYSON, R. B., MOORE, I. D. and MCMAHON, T. A., 1992a, Physically based hydrologic modeling. 1. A terrain-based model for investigative purposes, *Water Resources Research*, **28**, 2639-58.

GRAYSON, R. B., MOORE, I. D. and MCMAHON, T. A., 1992b, Physically based hydrologic modeling. 2. Is the concept realistic?, *Water Resources Research*, **28**, 2659-66.

GRIFFITH, D. A., 1988, *Advanced Spatial Statistics*, Dordrecht: Kluwer.

GRIFFITH, D. A., 1989, Distance calculations and errors in geographic databases, in GOODCHILD and GOPAL (1989), pp. 81-90.

GUARDIANO, F. B. and SRIVASTAVA, R. M., 1993, Multivariate geostatistics: beyond bivariate moments, in SOARES (1993), pp. 133-44.

GUPTILL, S. C. and MORRISON, J. L. (Eds.), 1995, *Elements of Spatial Data Quality*, Oxford: Pergamon.

HAINING, R. P., GRIFFITH, D. A. and BENNETT, R. J., 1983, Simulating two-dimensional autocorrelated surfaces, *Geographical Analysis*, **15**, 247-55.

HAMMERSLEY, J. M. and HANDSCOMB, D. C., 1979, *Monte Carlo Methods*, London: Chapman & Hall.

HARTS, J., OTTENS, H. F. L. and SCHOLTEN, H. J. (Eds.), 1991, *EGIS '91 Proceedings*, Utrecht: EGIS Foundation.

HEARNSHAW, H. M. and UNWIN, D. J. (Eds.), 1994, *Visualization in Geographical Information Systems*, Chichester: Wiley.

HELSTROM, C. W., 1991, *Probability and Stochastic Processes for Engineers*, New York: Macmillan.

HEUVELINK, G. B. M., 1992, An iterative method for multidimensional simulation with nearest neighbour models, in DOWD, P. A. and ROYER, J. J. (Eds.) *2nd CODATA Conference on Geomathematics and Geostatistics*, pp. 51–7, Nancy: Sciences de la Terre, Séries Informatique et Géologique 31.

HEUVELINK, G. B. M., 1996, Identification of field attribute error under different models of spatial variation, *International Journal of GIS*, **10**, 921–35.

HEUVELINK, G. B. M., 1998, Uncertainty analysis in environmental modelling under a change of spatial scale, *Nutrient Cycling in Agroecosystems* (Kluwer, in press).

HEUVELINK, G. B. M. and BIERKENS, M. F. P., 1992, Combining soil maps with interpolations from point observations to predict quantitative soil properties, *Geoderma*, **55**, 1–15.

HEUVELINK, G. B. M. and BURROUGH, P. A., 1993, Error propagation in cartographic modelling using Boolean logic and continuous classification, *International Journal of GIS*, **7**, 231–46.

HEUVELINK, G. B. M., BURROUGH, P. A. and LEENAERS, H., 1990, Error propagation in spatial modelling with GIS, in HARTS, J., OTTENS, H. F. L. and SCHOLTEN, H. J. (Eds.) *EGIS '90 Proceedings*, pp. 453–62, Utrecht: EGIS Foundation.

HEUVELINK, G. B. M., BURROUGH, P. A. and STEIN, A., 1989, Propagation of errors in spatial modelling with GIS, *International Journal of GIS*, **3**, 303–22.

HEUVELINK, G. B. M. and HUISMAN, J. A., 1996, Choosing between abrupt and gradual spatial variation?, in MOWRER, H. T., CZAPLEWSKI, R. L. and HAMRE, R. H. (Eds.) *Spatial Accuracy Assessment in Natural Resources and Environmental Sciences*, Fort Collins, Colorado: USDA Forest Service Technical Report RM-GTR-277, pp. 243–50.

HEUVELINK, G. B. M., MUSTERS, P. and PEBESMA, E. J., 1997, Spatio-temporal kriging of soil water content, in BAAFI and SCHOFIELD (1997), pp. 1020–30.

HUNTER, G. J. and GOODCHILD, M. F., 1996, A new model for handling vector data uncertainty in Geographic Information Systems, *URISA Journal*, **8**, 51–7.

HUNTER, G. J. and GOODCHILD, M. F., 1997, Modeling the uncertainty of slope and aspect estimates from spatial databases, *Geographical Analysis*, **29**, 35–49.

ISAAKS, E. H. and SRIVASTAVA, R. M., 1989, *An Introduction to Applied Geostatistics*, New York: Oxford University Press.

JANSEN, M. J. W., ROSSING, W. A. H. and DAAMEN, R. A., 1994, Monte Carlo estimation of uncertainty contributions from several independent multivariate sources, in GRASMAN, J. and VAN STRATEN, G. (Eds.) *Predictability and Nonlinear Modelling in Natural Sciences and Economics*, pp. 334–43, Dordrecht: Kluwer.

JANSSEN, P. H. M., HEUBERGER, P. S. C. and KLEPPER, O., 1994, UNCSAM: a tool for automating sensitivity and uncertainty analysis, *Environmental Software*, **9**, 1–11.

JOHNSON, M. E., 1987, *Multivariate Statistical Simulation*, New York: Wiley.

JONES, L., 1989, Some results comparing Monte Carlo simulation and first order Taylor series approximation for steady groundwater flow, *Stochastic Hydrology and Hydraulics*, **3**, 179–90.

JOURNEL, A. G., 1996, Modelling uncertainty and spatial dependence: stochastic imaging, *International Journal of GIS*, **10**, 517–22.

JOURNEL, A. G. and ALABERT, F., 1989, Non-Gaussian data expansion in the earth sciences, *Terra Nova*, **1**, 123–34.

JOURNEL, A. G. and ALABERT, F., 1990, New method for reservoir mapping, *Journal of Petroleum Technology*, **42**, 212–18.

JOURNEL, A. G. and HUIJBREGTS, CH. J., 1978, *Mining Geostatistics*, London: Academic Press.

KANDEL, A., 1986, *Fuzzy Mathematical Techniques with Applications*, Reading, Massachusetts: Addison-Wesley.

KEEFER, B. J., SMITH, J. L. and GREGOIRE, T. G., 1991, Modeling and evaluating the effects of stream mode digitizing errors on map variables, *Photogrammetric Engineering & Remote Sensing*, **57**, 957–63.

KERN, J. S., 1994, Spatial patterns of soil organic carbon in the contiguous United States, *Soil Science Society of America Journal*, **58**, 439–55.

KIIVERI, H. T., 1997, Assessing, representing and transmitting positional uncertainty in maps, *International Journal of GIS*, **11**, 33–52.

KING, P. R. and SMITH, P. J., 1988, Generation of correlated properties in heterogeneous porous media, *Mathematical Geology*, **20**, 863–77.

KITANIDIS, P. K., 1994, Generalized covariance functions in estimation, *Mathematical Geology*, **25**, 525–40.

KITANIDIS, P. K., 1997, A variance-ratio test for supporting a variable mean in kriging, *Mathematical Geology*, **29**, 335–48.

KLEMES, V., 1986, Dilettantism in hydrology: transition or destiny?, *Water Resources Research*, **22**, 177–88.

KLINKENBERG, B. and JOY, M., 1994, Visualizing uncertainty: succession or misclassification?, *Proceedings GIS/LIS*, pp. 494–503.

KLIR, G. J. and FOLGER, T. A., 1988, *Fuzzy Sets, Uncertainty and Information*, Englewood Cliffs, New Jersey: Prentice-Hall.

KOVAR, K. and NACHTNEBEL, H. P. (Eds.), 1993, *Application of Geographic Information Systems in Hydrology and Water Resources Management*, IAHS Publication No. 211, Wallingford: International Association of Hydrological Sciences Press.

KROS, J., DE VRIES, W., JANSSEN, P. H. M. and BAK, C. I., 1992, The uncertainty in forecasting trends of forest soil acidification, *Water, Air and Soil Pollution*, **66**, 29–58.

KUCZERA, G., 1988, On the validity of first-order prediction limits for conceptual hydrologic models, *Journal of Hydrology*, **103**, 229–47.

LANTER, D. P. and VEREGIN, H., 1992, A research paradigm for propagating error in layer-based GIS, *Photogrammetric Engineering & Remote Sensing*, **58**, 825–33.

LEENAERS, H., 1991, Deposition and storage of solid-bound heavy metals in the floodplains of the river Geul (The Netherlands), *Environmental Monitoring and Assessment*, **18**, 79–103.

LEENAERS, H., RANG, M. C. and RANG, D. M. C., 1991, Coping with uncertainty in the assessment of health risks, in FARMER, J. G. (Ed.), *Heavy Metals in the Environment*, pp. 286–9, Norwich: Page Bros.

LEENHARDT, D., 1995, Errors in the estimation of soil water properties and their propagation through a hydrological model, *Soil Use and Management*, **11**, 15–21.

LEMMENS, M. J. P. M., 1991, GIS: the data problem, in HARTS *et al.* (1991), pp. 626–36.

LEWIS, P. A. W. and ORAV, E. J., 1989, *Simulation Methodology for Statisticians, Operations Analysts, and Engineers*, Vol. 1, Pacific Grove, California: Wadsworth and Brooks/Cole.

LOAGUE, K., YOST, R. S., GREEN, R. E. and LIANG, T. C., 1989, Uncertainty in pesticide leaching assessment in Hawaii, *Journal of Contaminant Hydrology*, **4**, 139–61.

LODWICK, W. A., MONSON, W. and SVOBODA, L., 1990, Attribute error and sensitivity analysis of map operations in geographical information systems: suitability analysis, *International Journal of GIS*, **4**, 413–28.

LONGLEY, P., GOODCHILD, M. F., MAGUIRE, D. J. and RHIND, D. W. (Eds.), 1998, *Geographical Information Systems: Principles, Techniques, Management and Applications*, Cambridge: GeoInformation International.

LUIS, S. J. and MCLAUGHLIN, D., 1992, A stochastic approach to model validation, *Advances in Water Resources*, **15**, 15–32.

MACMILLAN, R. A., NIKIFORUK, W. L., KRZANOWSKI, R. M. and BALAKRISHNA, T. S., 1987, *Soil Survey of the Lacombe Research Station 1986 Expansion Area*, Vol. 1, Edmonton: Alberta Research Council.

MAFFINI, G., ARNO, M. and BITYTERLICH, W., 1989, Observations and comments on the generation of error in digital GIS data, in GOODCHILD and GOPAL (1989), pp. 55–67.

MAGUIRE, D. J., GOODCHILD, M. F. and RHIND, D. W. (Eds.), 1991, *Geographical Information Systems: Principles and Applications*, London: Longman.

MARDIA, K. V., 1972, *Statistics of Directional Data*, London: Academic Press.

MATÉRN, B., 1986, *Spatial Variation*, 2nd Edn, Lecture Notes in Statistics No. 36, Berlin: Springer.

MATHERON, G., 1973, The intrinsic random functions and their applications, *Advances in Applied Probability*, **5**, 439–68.

MATHERON, G., 1989, *Estimating and Choosing*, Berlin: Springer.

MCBRATNEY, A. B., WEBSTER, R. and BURGESS, T. M., 1981, The design of optimal sampling schemes for local estimation and mapping of regionalized variables: 1. Theory and method, *Computers and Geosciences*, **7**, 331–4.

MCDONALD, M. G. and HARBAUGH, A. W., 1984, A modular three-dimensional finite difference ground-water flow model, Report 83-875, Reston: US Geological Survey.

MEEKER, W. Q., CORNWELL, L. W. and AROIAN, L. A., 1980, The product of two normally distributed random variables, *Selected Tables in Mathematical Statistics* 7.

MEJIA, J. M. and RODRIGUEZ-ITURBE, I., 1974, On the synthesis of random field sampling from the spectrum: an application to the generation of hydrologic spatial processes, *Water Resources Research*, **10**, 705–11.

MOORE, R. D. and ROWLAND, J. D., 1990, Evaluation of model performance when the observed data are subject to error, *Physical Geography*, **11**, 379–92.

MULLER, J. C., 1987, The concept of error in cartography, *Cartographica*, **24**, 1–15.

MYERS, D. E., 1982, Matrix formulation of co-kriging, *Mathematical Geology*, **14**,

249–57.

NETHERLANDS SOIL SURVEY INSTITUTE, 1975, Toelichting bij de bodemkaart van Nederland, blad (40 O-W), schaal 1 : 50 000, Wageningen (in Dutch).

NEWCOMER, J. A. and SZAJGIN, J., 1984, Accumulation of thematic map errors in digital overlay analysis, *The American Cartographer*, **11**, 58–62.

OBERTHÜR, T., DOBERMANN, A. and NEUE, H. U., 1996, How good is a reconnaisance soil map for agronomic purposes?, *Soil Use and Management*, **12**, 33–43.

OEHLERT, G. W., 1992, A note on the delta method, *American Statistician*, **46**, 27–9.

OKX, J. P., HEUVELINK, G. B. M. and GRINWIN, A. W., 1990, Expert knowledge and (geo)statistical methods: complementary tools in soil pollution research, in ARENDT, F., HISENVELD, M. and VAN DEN BRINK, W. J. (Eds.), *Contaminated Soil '90*, pp. 729–36, Dordrecht: Kluwer.

OLIVER, M. A. and WEBSTER, R., 1990, Kriging: a method of interpolation for geographical information systems, *International Journal of GIS*, **4**, 313–32.

OPENSHAW, S., 1989, Learning to live with errors in spatial databases, in GOODCHILD and GOPAL (1989), pp. 263–76.

OPENSHAW, S., CHARLTON, M. and CARVER, S., 1991, Error propagation: a Monte Carlo simulation, in MASSER, I. and BLAKEMORE, M. (Eds.), *Handling Geographical Information*, pp. 78–101, Harlow, United Kingdom: Longman.

PAPRITZ, A. and WEBSTER, R., 1995, Estimating temporal change in soil monitoring: II. Sampling from simulated fields, *European Journal of Soil Science*, **46**, 13–27.

PARRATT, L. G., 1961, *Probability and Experimental Errors in Science*, New York: Wiley.

PARZEN, E., 1962, *Stochastic Processes*, San Francisco: Holden-Day.

PEBESMA, E. J. and WESSELING, C. G., 1998, Gstat, a program for geostatistical modelling, prediction and simulation, *Computers and Geosciences* (IAMG, in press).

PECK, A. J., GORELICK, S. M., DE MARSILY, G., FOSTER, S. and KOVALEVSKY, V. S., 1988, *Consequences of Spatial Variability in Aquifer Properties and Data Limitations for Groundwater Modelling Practice*, IAHS Publication No. 175, Wallingford: International Association of Hydrological Sciences Press.

PRESS, W. H., TEUKOLSKY, S. A., VETTERLING, W. T. and FLANNERY, B. P., 1992, *Numerical Recipes in C*, 2nd Edn, Cambridge: Cambridge University Press.

RALL, L. B., 1981, *Automatic Differentiation: Techniques and Application*, Berlin: Springer.

RIPLEY, B. D., 1987, *Stochastic Simulation*, New York: Wiley.

ROSENBLUETH, E., 1975, Point estimates for probability moments, *Proceedings of the National Academy of Sciences of the United States of America*, **72**, 3812–14.

ROSSI, R. E., BORTH, P. W. and TOLLEFSON, J. J., 1993, Stochastic simulation for characterizing ecological spatial patterns and appraising risk, *Ecological Applications*, **3**, 719–35.

SCAVIA, D., POWERS, W. F., CANALE, R. P. and MOODY, J. L., 1981, Comparison of first-order error analysis and Monte Carlo simulation in time-dependent lake eutrophication models, *Water Resources Research*, **17**, 1051–9.

SCHUIT, S., 1989, Interpolatie van ruimtelijke gegevens met behulp van het Nearest Neighborhood Model, Master's thesis, Twente Technical University, The Netherlands (in Dutch).

SCHWEPPE, F. C., 1973, *Uncertain Dynamical Systems*, Englewood Cliffs, New Jersey: Prentice-Hall.

SHARP, W. E. and AROIAN, L. A., 1985, The generation of multidimensional autoregressive series by the herringbone method, *Mathematical Geology*, **17**, 67–79.

SKIDMORE, A. K., 1989, A comparison of techniques for calculating gradient and aspect from a gridded digital elevation model, *International Journal of GIS*, **3**, 323–34.

SMITH, L. and FREEZE, R. A., 1979, Stochastic analysis of steady state groundwater flow in a bounded domain. 2. Two-dimensional simulations, *Water Resources Research*, **15**, 1543–59.

SMITH, R. E. and HEBBERT, R. H. B., 1979, A Monte Carlo analysis of the hydrologic effects of spatial variability of infiltration, *Water Resources Research*, **15**, 419–29.

SNEDECOR, G. W. and COCHRAN, W. G., 1989, *Statistical Methods*, 8th Edn, Ames: Iowa State University Press.

SOARES, A. (Ed.), 1993, *Geostatistics Tróia '92*, Dordrecht: Kluwer.

SOIL SURVEY STAFF, 1975, *Soil Taxonomy; A Basic System of Soil Classification for Making and Interpreting Soil Surveys*, Agriculture Handbook No. 436, Department of Agriculture, Washington, DC.

STANISLAWSKI, L. V., DEWITT, B. A. and SHRESTHA, R. L., 1996, Estimating positional accuracy of data layers within a GIS through error propagation, *Photogrammetric Engineering & Remote Sensing*, **62**, 429–33.

STARKS, T. H., 1986, Determination of support in soil sampling, *Mathematical Geology*, **18**, 529–37.

STEHMAN, S. V., 1992, Comparison of systematic and random sampling for estimating the accuracy of maps generated from remotely sensed data, *Photogrammetric Engineering & Remote Sensing*, **58**, 1343–50.

STEIN, A., BOUMA, J., MULDERS, M. A. and WETERINGS, M. H. W., 1989, Using spatial variability studies to estimate physical land qualities of a level river terrace, *Soil Technology*, **2**, 385–402.

STEIN, A., HOOGERWERF, M. and BOUMA, J., 1988, Use of soil-map delineations to improve (co-)kriging of point data on moisture deficits, *Geoderma*, **43**, 163–77.

STEIN, M., 1987, Large sample properties of simulations using Latin hypercube sampling, *Technometrics*, **29**, 143–51.

TAYLOR, J. R., 1982, *An Introduction to Error Analysis*, Mill Valley: University Science Books.

TEN CATE, B., BRUS, D. J. and DE GRUIJTER, J. J., 1997, Random sampling or geostatistical modelling? Choosing between design-based and model-based sampling strategies for soil (with Discussion), *Geoderma*, **80**, 1–60.

THAPA, K. and BOSSLER, J., 1992, Accuracy of spatial data used in geographic information systems, *Photogrammetric Engineering & Remote Sensing*, **58**, 835–41.

THEOBALD, D. M., 1989, Accuracy and bias issues in surface representation, in GOODCHILD and GOPAL (1989), pp. 99–106.

TIETEMA, A. and VERSTRATEN, J. M., 1992, Nitrogen cycling in an acid forest ecosystem in the Netherlands under increased atmospheric nitrogen input: the nitrogen budget and the effect of nitrogen transformations on the proton budget, *Biogeochemistry*, **15**, 21–46.

TOMLIN, C. D., 1983, Digital cartographic modelling techniques in environmental planning, PhD thesis, Yale University, Connecticut.

TOMLIN, C. D., 1990, *Geographic Information Systems and Cartographic Modeling*, Englewood Cliffs, New Jersey: Prentice-Hall.

TOPPING, J., 1962, *Errors of Observation and their Treatment*, London: Chapman and Hall.

UPTON, G. J. G. and FINGLETON, B., 1989, *Spatial Data Analysis by Example*, Chichester: Wiley.

VAN DER PERK, M., 1997, Effect of model structure on the accuracy and uncertainty of results from water quality models, *Hydrological Processes*, **11**, 227–39.

VAN DER SLUIJS, P. and DE GRUIJTER, J. J., 1985, Water table classes: a method to describe seasonal fluctuation and duration of water tables on Dutch soil maps, *Agricultural Water Management*, **10**, 109–25.

VAN DEURSEN, W. P. A., 1995, Geographical information systems and dynamic models: development and application of a prototype spatial modelling language, PhD thesis, University of Utrecht, The Netherlands.

VAN DEURSEN, W. P. A. and WESSELING, C. G., 1995, PCRaster Software, Department of Physical Geography, University of Utrecht, http://www.frw.ruu.nl/pcraster.html.

VAN DIEPEN, C. A., VAN KEULEN, H., WOLF, J. and BERKHOUT, J. A. A., 1991, Land evaluation: from intuition to quantification, *Advances in Soil Science*, **15**, 139–204.

VAN DIEPEN, C. A., WOLF, J., VAN KEULEN, H. and RAPOLDT, C., 1989, WOFOST: a simulation model of crop production, *Soil Use and Management*, **5**, 16–24.

VAN GEER, F. C., TE STROET, C. B. M. and YANGXIAO, Z., 1991, Using Kalman filtering to improve and quantify the uncertainty of numerical groundwater simulations. 1. The role of system noise and its calibration, *Water Resources Research*, **27**, 1987–94.

VAN KUILENBURG, J., DE GRUIJTER, J. J., MARSMAN, B. A. and BOUMA, J., 1982, Accuracy of spatial interpolation between point data on soil moisture supply capacity, compared with estimates from mapping units, *Geoderma*, **27**, 311–25.

VAN LANEN, H. A. J., VAN DIEPEN, C. A., REINDS, G. J., DE KONING, G. H. K., BULENS, J. D. and BREGT, A. K., 1992, Physical land evaluation methods and GIS to explore the crop growth potential and its effects within the European Communities, *Agricultural Systems*, **39**, 307–28.

VANMARCKE, E., 1983, *Random Fields: Analysis and Synthesis*, Cambridge, Massachuset: MIT Press.

VAN WIJNEN, J. H., CLAUSING, P. and BRUNEKREEF, B., 1990, Estimated soil ingestion by children, *Environmental Research*, **51**, 147–62.

VEREECKEN, H., MAES, J., FEYEN, J. and DARIUS, P., 1989, Estimating the soil moisture retention characteristic from texture, bulk density and carbon content, *Soil Science*, **148**, 389–403.

VEREGIN, H., 1989a, A taxonomy of error in spatial databases, Technical paper 89-12, National Center for Geographic Information and Analysis, Santa Barbara, California.

VEREGIN, H., 1989b, Error modeling for the map overlay operation, in GOODCHILD and GOPAL (1989), pp. 3–18.

VOLTZ, M. and WEBSTER, R., 1990, A comparison of kriging, cubic splines and classification for predicting soil properties from sample information, *Journal of Soil Science*, **41**, 473–90.

WAGNER, B. J. and GORELICK, S. M., 1987, Optimal groundwater quality management under parameter uncertainty, *Water Resources Research*, **23**, 1162–74.

WEBSTER, R. and OLIVER, M. A., 1990, *Statistical Methods in Soil and Land Resource Survey*, Oxford: Oxford University Press.

WESSELING, C. G. and HEUVELINK, G. B. M., 1991, Semi-automatic evaluation of error propagation in GIS operations, in HARTS et al. (1991), pp. 1228–37.

WESSELING, C. G. and HEUVELINK, G. B. M., 1993a, ADAM User's Manual, Department of Physical Geography, University of Utrecht, The Netherlands.

WESSELING, C. G. and HEUVELINK, G. B. M., 1993b, Manipulating quantitative attribute accuracy in vector GIS, in HARTS, J., OTTENS, H. F. L. and SCHOLTEN, H. J. (Eds) *Proceedings EGIS '93*, pp. 675–84, Utrecht: EGIS Foundation.

WESSELING, C. G., KARSSENBERG, D., BURROUGH, P. A. and VAN DEURSEN, W. P. A., 1996, Integrating dynamic environmental models in GIS: the development of a dynamic modelling language, *Transactions in GIS*, **1**, 40–8.

WETERINGS, M. H. W., 1988, Variabiliteit van een landkarakteristiek en van gesimuleerde gewasopbrengsten binnen een rivierterras van de Allier in de Limagneslenk, Frankrijk, Student's Report of the Department of Soil Science and Geology, Agricultural University of Wageningen, The Netherlands (in Dutch).

WHITTLE, P., 1954, On stationary processes in the plane, *Biometrika*, **41**, 434–49.

WILKS, S. S., 1962, *Mathematical Statistics*, New York: Wiley.

WILLMOTT, C. J., 1981, On the validation of models, *Physical Geography*, **2**, 184–94.

WINGLE, W. L., POETER, E. P. and MCKENNA, S. A., 1996, UNCERT: a geostatistical uncertainty analysis package applied to groundwater and contaminant transport modeling, http://www.mines.edu/fs_home/wwingle/uncert/.

WOLDT, W., GODERYA, F., DAHAB, M. and BOGARDI, I., 1996, Consideration of spatial variability in the management of non-point source pollution to groundwater, in MOWRER, H. T., CZAPLEWSKI, R. L. and HAMRE, R. H. (Eds.) *Spatial Accuracy Assessment in Natural Resources and Environmental Sciences*, Fort Collins, Colorado: USDA Forest Service Technical Report RM-GTR-277, pp. 49–56.

WÖSTEN, J. H. M., BOUMA, J. and STOFFELSEN, G. H., 1985, The use of soil survey data for regional soil water simulation models, *Soil Science Society of America Journal*, **49**, 1238–44.

YOST, R. S., LOAGUE, K. and GREEN, R. E., 1993, Reducing variance in soil organic carbon estimates: soil classification and geostatistical approaches, *Geoderma*, **57**, 247–62.

ZHANG, R. and YANG, J., 1996, Iterative solution of a stochastic differential equation: an efficient method for simulating soil variability, *Geoderma*, **72**, 75–88.

Author index

Subject index

ADAM
 application of the ADAM error
  propagation tool 93–5
 comparison with other tools 95
 description of the tool 89–93, 102, 107
anisotropy 15
approximation error
 in stochastic spatial simulation 79
 with Rosenblueth's method 54, 59
 with the Taylor methods 36–8, 43–4, 95,
  99, 105
 attribute error 6, 10, 12–13, 49, 97, 103
autocorrelation
 of a DEM 57, 59–60, 100
 of a model of spatial variation 14–15,
  18–20, 31
 of autoregressive random fields 87, 105
 of the error model 11, 35, 46, 97
 of the output of an error analysis 106
autocovariance 11, 15, 18–20, 81
autoregressive random field 80–2, 84–6, 105

balance of errors 48–9, 102
bias
 of attribute error 10, 34, 97
 unbiasedness in Monte Carlo estimation
  41–2
 unbiasedness in spatial prediction 17–18,
  20–1, 29
block kriging, *see* kriging
Boolean classification 67–76, 100, 105

cartographic algebra 2–3
categorical error 6
change of support 22–3, 55, 66, 79, 103
Cholesky decomposition 40, 79, 82
coefficient of variation 41, 91
cokriging, *see* kriging

computational load 43, 45, 81, 90, 95, 105
computational model
 definition 2–3, 10
 error of a computational model 47, 49, 106,
  108
 error propagation in computational
  models 5–6, 43, 45, 90, 101
conceptual model 3, 6, 44, 48, 80–1, 101
conditional simulation 79
continuous classification 67–8, 71–6, 92, 100,
 105
continuous model of spatial variation
 (CMSV)
 application of the CMSV 26–32
 comparison with other models of spatial
  variation 14, 16–17, 23, 107
 definition and identification of the CMSV
  15, 18–19, 98
 multivariate extension of the CMSV 20–1
contribution of error sources
 relative contribution of input and model
  error 49, 65–66, 102, 104
 relative contribution of multiple input
  errors 46–7, 49, 92, 101–2, 104
cross-correlation 20, 35, 90–1

design-based approach 23–4
digital elevation model (DEM)
 error analysis with DEMs 4, 13, 56–7, 60,
  100
 use of DEMs in GIS 2, 77, 92
discrete model of spatial variation (DMSV)
 application of the DMSV 26–32
 comparison with other models of spatial
  variation 16–17, 23, 103, 107
 definition and identification of the DMSV
  14–15, 18, 20, 98

empirical model 3, 6, 44, 48

For Product Safety Concerns and Information please contact our EU
representative GPSR@taylorandfrancis.com
Taylor & Francis Verlag GmbH, Kaufingerstraße 24, 80331 München, Germany

www.ingramcontent.com/pod-product-compliance
Ingram Content Group UK Ltd.
Pitfield, Milton Keynes, MK11 3LW, UK
UKHW051942210425
457613UK00026BA/191